MECHANICS, HEAT, AND THE HUMAN BODY

HOWARD D. GOLDICK

University of Hartford

Library of Congress Cataloging-in-Publication Data

Goldick, Howard D.
  Mechanics, heat, and the human body. Howard D. Goldick.
    p.cm.
  Includes bibliographical references and index.
  ISBN 0-13-922816-0
    1. Biophysics. 2. Physics. 3. Human mechanics. 4. Mechanics.
  5. Heat. 6. Thermodynamics. I. Title.

QH505 .G63 2000
612'.014--dc21   00-044586

Publisher: Julie Alexander
Acquisitions Editor: Mark Cohen
**Director of Production
  and Manufacturing:** Bruce Johnson
Managing Production Editor: Patrick Walsh
Production Editor: Carol Eckhart,
  York Production Services
Production Liaison: Larry Hayden
Manufacturing Manager: Tiffany Price
Creative Director: Marianne Frasco
Cover Design Coordinator: Maria Guglielmo
Director of Marketing: Leslie Cavaliere
Marketing Coordinator: Cindy Frederick
Editorial Assistant: Melissa Kerian
Composition: York Production Services
Printing and Binding: R.R. Donnelley

Prentice-Hall International (UK) Limited, *London*
Prentice-Hall of Australia Pty. Limited, *Sydney*
Prentice-Hall Canada Inc., *Toronto*
Prentice-Hall Hispanoamericana, S.A., *Mexico*
Prentice-Hall of India Private Limited, *New Delhi*
Prentice-Hall of Japan, Inc., *Tokyo*
Prentice-Hall Singapore Pte. Ltd.
Editora Prentice-Hall do Brasil, Ltda., *Rio de Janeiro*

Photography and Artwork Credits:

Fishbane PM, Gasiorowicz S, Thornton ST,
Physics for Scientists and Engineers,
2nd ed, 1996, Prentice-Hall, (f 1.3) 8,
(f 1.8) 25, (f 1.61) 75, (f 2.1A) 90, (f 2.2)
91, (f 2.7) 97, (f 2.33) 118, (f 3.7) 162,
(f 3.8) 162. • **Martini, FH, Timmons, MJ,**
Human Anatomy, **2nd ed, 1997, Prentice-
Hall,** (f 1.4) 11, (f 1.29) 39, (f 1.43A) 56,
(f 3.17) 185, (A 2) 201, (A 3) 202. • **Giancoli
DC,** *Physics for Scientists and Engineers,*
3rd ed, 2000, Prentice-Hall, (f 1.55) 69,
(f 2.1B-C) 90, (f 3.9) 164, (f 3.10) 165.
• **Discover Magazine, July 1995,** (f 3.4) 149.

$43.00 1110731

Prentice
Hall

10 9 8 7 6 5 4 3 2 1
ISBN 0-13 922816-0

CONTENTS

ACKNOWLEDGMENTS

When I began writing the class notes on which this book is based, I had no idea that it would result in writing a textbook. Having no prior experience in writing a text, I needed a great deal of support and encouragement. In particular, I needed expressions of encouragement in what was a long, time-consuming effort. I want to thank the chairman of my department, Joel Kagan, for his initial and continuing encouragement and support. When I began to search for a publisher, I found that there was little or no interest in my concept of the text. Sandi Hakanson, who represents Prentice Hall, encouraged me to submit my book to Mark Cohen, Health Sciences Acquisitions Editor for Prentice Hall. He encouraged me to submit a proposal and then arranged a contract. His continuing encouragement and support made the completion of the text possible. There were several people who read the text and made extremely helpful comments. Nancy Beverly of Mercy College used a preliminary version in one of her classes. David Markowitz of the University of Connecticut read two drafts and gave me many detailed comments. Howard Goldberg of the University of Hartford also read the final draft and helped me to adjust my particular way of writing to a more widely accepted style. I wish to thank Scott Sechrest of the CYBEX Corporation and Robert B. Barnes of the Barnes Engineering Corporation for their encouragement and permission to use proprietary illustrations. Finally, I thankfully acknowledge the many students who worked with the early versions of the notes on which this book is based. Without their comments and suggestions, I would never have continued in this effort.

PREFACE

In writing this book, I had the goal of providing an introduction to physics for those students who are particularly interested in the human body. On the basis of my many years of teaching physics to premedical, physical therapy and occupational therapy students, I set these guidelines:

1. The text would cover a limited number of distinct physics topics rather than providing an encyclopedic survey of the field of physics.
2. These topics would be illustrated (examples and problems) with reference to specific functions and characteristics of the human body.
3. The material would be covered in greater depth than is typical of an introductory text. This provides an opportunity to demonstrate the roles that physics and mathematical analysis play in understanding the body.
4. The examples and problems would span a range from straightforward applications of basic physics principles to those requiring significant analysis.

My students have, during the past five years, used the notes on which the text is based as a stand-alone text for a one-semester course. Much of the present content is based on their questions, criticisms, and suggestions. For example:

1. The discussion of each topic is built around a series of steps on which the analysis is based.
2. Both the SI and USA (English) systems of units are used in the book. Although the SI system is the legal system in this country and is the most commonly used system in the sciences, it is not widely used outside of those fields. Therefore, most students are much more familiar with the USA system, and this familiarity is addressed by inclusion of the USA system.
3. The various tables indicate sources of the data in the bibliography.
4. Answers to all of the quantitative problems are included.

I strongly suggest that students who use this book do not limit their efforts to reading it. To derive the full benefits that I hope are present, it is necessary that during your reading, you fill in any gaps between equations. There should be no "magic," no material that seems to come from nowhere. Do as many problems as your time allows. In your analyses of these problems, follow the suggested procedures rather than using shortcuts. Each analysis should include the basic applicable physics principle and clearly show how it is used. The answers to all problems are given. Do not work from these answers backward to produce your analysis. Such an approach is self-defeating because you will not be given the answers on exams or if you enter a field in which you must carry out such analyses.

One last comment: This text is intended to be a physics book, not an anatomy or physiology text. The human body is extremely complex, and to deal with its functions at an introductory level, many simplifications have been made. Modeling is employed; for example, muscles are treated as if they are simple line forces. Nevertheless, the results of the analyses are illustrative of the body's functions.

I look forward to your comments and questions regarding the book. Please contact me via e-mail at goldick@mail.hartford.edu.

INTRODUCTION

Our understanding of the human body and the means by which we deal with maladies and injuries have undergone amazing changes during the last 100 years. Illnesses that had been viewed as the result of Divine Intervention are now viewed in terms of the effects of bacteria and/or viruses. Amputation was a common medical response to severe trauma to limbs but is now very rare. The field of prosthetics has advanced to such a degree that those who have lost limbs are no longer doomed to living a marginal life but may now lead so full a life that it is sometimes difficult to realize that they have such a handicap. In the past, a person who had suffered a spinal cord injury that resulted in loss of the use of his or her legs could look forward only to life in a wheelchair. A person suffering that injury today can reasonably hope to walk and even climb stairs. Whereas exploratory surgery was common in the past, it is now very rare, having been replaced by noninvasive means. These and many other medical advances testify to the central role that the physical sciences and technology play in our dealings with the human body.

In this text, we will deal with the application of certain aspects of physics (mechanics and heat) to the human body. We will answer questions such as the following:

1. If a 150-pound woman were standing while holding a 10-pound child, how much force would be acting to compressing her lower back? (About 109 pounds) She bends over to put the child down into a playpen. How much force is now compressing her lower back? (439 pounds) (See page 109.)

2. Why does a person who has injured his right hip lean toward his right when walking? Why should he use a cane on his left rather than his right side? (See pages 113-115.)

3. What is the average power output of a catcher while stopping a fastball? (5 hp) (See page 146.)

4. How many times would you have to curl an 11-pound weight to burn off the energy you take in by eating six chocolate chip cookies? (5000) (See page 171.)

5. You know that your body produces heat when you exercise. How does the rate at which your body produces heat compare to the rate at which a 100-watt light-bulb produces heat? Surprisingly, even when you are not exerting yourself, as

while lying still in bed, you are producing heat at a rate comparable to that of the lightbulb. (See page 148.)

6. Why does your body seems to produce and retain fat so easily, and why is it so difficult to lose the fat? (See page 150.)

7. What is the function of kneecaps? (See page 192.)

8. Why is your spinal column curved rather than straight? (See page 97.)

9. Why does a pregnant woman usually lean backward when standing? (See page 97.)

10. How is it possible for a cold-blooded animal such as a tuna or a shark to have an internal temperature that is higher than that of the cold water in which it swims? (See page 188.)

As we learn how to analyze these and many other situations, we will become familiar with concepts that are basic to physics, such as Newton's laws and conservation of energy. We will also learn about the anatomy and physiology of the human body; in particular, we will deal with the muscular-skeletal system, digestion, and temperature regulation systems.

Perhaps more important than this information, which can be found in many books, is the techniques of analysis and quantitative reasoning that we will develop. **In my opinion, it has been the application of these techniques that has made possible the amazing advances in medicine and health care in general that we enjoy today.**

HISTORICAL BACKGROUND

Our efforts to understand or explain the world seem to be inherent. Evidence for this statement comes from such diverse areas of study as comparative mythology and child psychology. Just as a child repetitively asks "Why?" and seems never to be satisfied by the answers, so it was with our ancestors. Unfortunately, this attitude is not supported by contemporary culture and has been replaced by a sort of sophistication and noncritical collective agreement characterized by

"OK?"

"Sure."

Try to imagine a culture where the interchange would be

"OK?"

"No, explain it more clearly."

That is the culture that our studies will represent.

Our studies will deal with the human body. How do we come to understand the body? This has been a long and difficult process. As we shall see, there were many questions that we would consider to be perfectly legitimate but that were, for many hundreds of years, the province of religion rather than science. There are many obvious questions about the body that must have been raised recurrently since time immemorial. Such questions as

"Where do babies come from?"

"Why do people die?"

"How can I get rid of this cold?"

"How can I get rid of this headache?"

are ancient; many of them have only recently been answered, and some of them still do not have definitive answers.

I remember attending an exhibit of cave art at the Metropolitan Museum of Art in New York City several years ago. The exhibit consisted of artifacts and reproductions of drawings that had been found in caves in France and Spain. These represented the artistic accomplishments of people who lived approximately 40,000 years ago. As I examined the exhibit, I noticed that although there were many female fertility symbols—small statues of female figures emphasizing breasts, hips, and bellies—there were no male fertility symbols, that is, phallic symbols. I asked an attendant whether the exclusion had been purposeful and was directed to an animal's horn that had been decorated with drawings. Still curious about the relative abundance of female symbols and scarcity of male symbols, it came to me that perhaps these artifacts dated from before the time when people realized that the male had anything to do with making babies. The role of the female is obvious, but who could remember and associate with the birth an activity that had happened nine months earlier?

No wonder that conception and birth were viewed as mysterious events, playing a central role in mythology and religion. As with conception and birth, so too with disease and death. It seemed to early people that one could divide concerns about the body into two categories: those that were inherently mysterious, such as conception, death, and disease, and those that were directly observable and hence understandable, such as wounds. It became accepted that while the former were to be dealt with through religion and other spiritual—that is, nonphysical—means (see the Book of Job), the latter situations were amenable to human intervention, such as stopping bleeding and setting broken bones. This separation was widely accepted through the eighteenth century.

Today, most people accept the physical, as opposed to spiritual, bases of birth, death, and disease. This change has affected human perception to such a degree that when there is no quick, effective intervention for such maladies as the common cold and AIDS, some people find it easier to believe that there is a conspiracy rather than a lack of scientific knowledge. How did such a massive change in attitude come about?

The major transition seems to have occurred in Europe in the seventeenth and eighteenth centuries, during what has been called the Scientific Revolution. Before that time, the generally accepted way to find answers to questions about the world was to look in old books. These books were usually Latin translations of the works of people who had lived between 300 B.C. and A.D. 200 in the world associated with classical Greece and the Hellenistic period that followed it. During the Roman period, this knowledge was disseminated throughout the empire by traveling scholars and physicians. With the fall of the empire in the fifth century, the flow of information ceased, and the so-called Dark Ages in Europe began. The knowledge was not lost, however. Some books had been kept and studied in monasteries. Other Europeans became aware of these works while engaging in commercial contacts with the Islamic world and as a result of the Crusades.

Many of the books that had originally been written in Greek and Latin were translated into Arabic after the Islamic conquests (632–750) of the Hellenistic cities of Spain, North Africa, and the Middle East. There were several centers of translation in the Islamic world.

Chief among them were ninth century Baghdad, tenth century Cairo, and twelfth century Toledo in Spain. Among the works translated were those of Aristotle, Archimedes, Hippocrates, Galen, and Euclid. The Hellenistic tradition of study in the fields of medicine, physics, and mathematics was continued by the Moslems, and this accumulated knowledge gradually began to find its way back into Europe. Many more translations became available to scholars in Europe after the Christian conquest of Islamic Spain in 1492. The impact that these works had on Europeans cannot be overestimated. They were viewed as the works of the people who had built the cities of Rome, Athens, Constantinople, and Alexandria, all of which were far more impressive than the largest cities of Europe. Thus, they were taken as Truth and as the source of all true knowledge about the physical world.

Of the Greek and Hellenistic scholars mentioned above, Claudius Galen (c. 130–200) deserves our particular attention. About a hundred of his works became available to Europeans through the processes described above. He was born in Pergamum (a large Hellenistic city located in what is now Turkey) and studied medicine there and at other major centers of Hellenistic learning in Smyrna, Corinth, and Alexandria. He then returned to Pergamum, where he served as a surgeon to a school for gladiators. He later went to Rome, where he became the physician to Emperor Marcus Aurelius. During his lifetime, he wrote many books, not only describing his investigations in the field of medicine but also recording the beliefs of others. His writings on anatomy were partly based on his surgical experience with the gladiators and on his experiences while accompanying the emperor on campaigns against German tribes. However, most of his knowledge of anatomy was based on dissections that he carried out on Barbary apes. His writings on physiology were based on observations but also on the prevailing philosophical traditions of his times. He believed that the functions of the body were based on spirits that endow us with the abilities to grow (natural spirit), to move (vital spirit), and to think (animal spirit). These spirits were not religious, supernatural, or mystical but were derived from the air in the liver, the heart, and the brain, respectively. His use of the word "spirits" was similar to our use of the word when we refer to ammonia spirits or spirits of alcohol—more as a vapor than a ghost. Galen's work was taken as gospel by the physicians of the Middle Ages, not to be questioned or subjected to independent verification. This attitude began to diminish with the work of Andreas Vesalius (1515–1564).

His major work, *De Humani Corporis Fabrica* (*About the Workings of the Human Body*, usually referred to as *Fabrica*), published in 1543, did not merely repeat what Galen had written. Vesalius described what he had observed while doing dissection. However, he accepted the main ideas about physiology that Galen had propounded. His contributions to anatomy that clearly showed errors in the Galenic texts set the stage for continuing investigations.

In 1687, Isaac Newton published *Philosophiae Naturalis Principia Mathematica* (*The Mathematical Principles of Natural Philosophy*, usually referred to as the *Principia*). In this book, Newton argued that the physical world could be understood, not by reference to the old books but by the application of close observation and logical analysis employing mathematics. He demonstrated the usefulness of this approach by showing how he could explain the motions of the planets, the comets, and the moon and the cause of high and low tides. His work made a great impression on those who were trying to understand the world and were not satisfied with the old answers. Although Newton's work emphasized the fields that we would call physics and astronomy, his work also made an impact on medicine.

In 1702, Richard Mead, a physician, published *A Mathematical Account of Poisons*. The book began with the claim that the study of mathematics would show doctors how to solve the intractable problems of medicine. His contemporary, Giorgio Baglivi, professor of anatomy in Rome wrote (paraphrased), "the human body is truly nothing else but a complex of chemical-mechanical motions, depending on such principles as are purely mathematical. For whoever takes an attentive view of its fabric, he'll really meet with shears in the jaw-bones and teeth . . . a pair of bellows in the lungs, the power of a lever in the muscles, pulleys in the corners of the eyes, and so on." (Lenihan, 1975.) Later in the book, Baglivi wrote, "We must not be surprised to find that the true and genuine cause of diseases can never be found by theoretical philosophical principles."

Even with the appearance of such innovative works, the Galenic ideas relating disease to humors and spirits continued to be widely held. The removal of such spirits from a sick person by techniques such as blood letting and trepanning could be found well into the nineteenth century.

The idea that one could use physics and mathematics to better understand the body was wonderfully supported by the discovery of X-rays, announced by Professor W. C. Roentgen of the University of Wurzenberg in Germany in November 1895. The medical applications of this new phenomenon followed with amazing speed. The first medical X-ray in the United States was taken on February 3, 1896, by Professor Edwin Frost of Dartmouth. In March 1896, Dr. J. Daniels of Vanderbilt University announced that irradiation of a colleague's skull had resulted in hair loss. Removal of a hairy birthmark was reported in 1897 and of a skin tumor in 1899. The success of X-rays established the great utility of physics and mathematics in the efforts to understand the human body.

Since the turn of the twentieth century, the most commonly accepted mode of gaining understanding of the body has been characterized by the application of biology, chemistry, physics, and mathematics. The almost magical level of medical technology is built on these bases. We will now begin a detailed study of this process.

1

LINEAR MOTION AND FORCE

One of the most obvious characteristics of the body is that of motion. Not only did people wonder what makes the body move but they even debated the very nature of motion. One of the recurring topics in mythology and religion has been to explain the motion of the sun, the stars, and the moon. On a much more mundane level, it is easily noticed that some things in our environment move and some do not. Of those that do move, some move because something else (such as the wind) makes them move, while others move on their own. The things that move on their own are generally considered to be alive; we call them animals. Notice the similarity between the word "animal" and the words "animate" and "animation." These words come from a Greek word for breath, soul, or spirit. It came to mean "alive". Thus an animated picture is one that seems to be alive; it is a moving picture. We realize that plants are also alive, but since they do not move on their own, they are somehow considered less alive than are animals. Other things, such as rocks, do not move unless something makes them move, and they are clearly not alive. So we can consider two categories: things such as animals that can move on their own and all other things.

One of the earliest writers on this subject was Aristotle (384–322 B.C.). The effort to understand motion played a major role in his analysis of the physical world. He wrote that each inanimate (nonanimal) object has a natural, or inherent, motion that is either straight

1

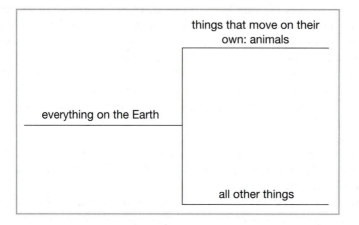

up or straight down, depending on the major constituent component of the object. The four basic substances of which everything is composed were fire, air, water and earth. (*Note*: The capitalized terms are used to distinguish between normal fire, air, water, and earth and the Aristotelian substances of which normal materials are formed.) According to Aristotle, the natural motion of fire and of air is straight up, and that of water, and of earth is straight down. Natural motion was inherent in the object, thus not requiring any external agent. An object that displayed any motion other than its natural motion was said to be exhibiting unnatural or violent motion. **If an object is observed to be moving in a direction that is not consistent with its natural motion then an external agent or force must be at work causing the unnatural or violent motion.**

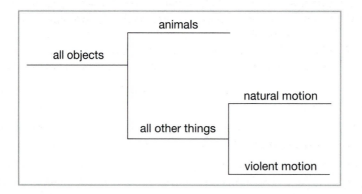

For example, if a rock is seen to be moving straight down, it is not surprising because the rock is made of earth and so its natural motion is straight down. But if the rock were observed to be moving up or horizontally, then there must be some force or external agent acting on the rock, causing it to move with the observed unnatural motion. Maybe the rock is in someone's hand, or maybe a very strong wind is blowing. When water mixes with air (evaporation) or comes out of a fountain, it goes up, but this is not the natural motion of water. Eventually, its natural motion must take over, and it falls (rain). If embers or dust rise because they are affected by hot air from a fire, they will eventually fall back because of their natural motion.

Since most of the motion that we observe going on around us is neither straight up nor straight down, it may be said that most motion is caused by external forces. This concept was generally accepted from the fourth century B.C. until the early seventeenth century, when Galileo Galilei (1564–1642) successfully challenged it.

Galileo's two major works were the books *Dialogue Concerning the Two Chief World Systems (1632) and Discourses and Mathematical Demonstrations Concerning Two New Sciences Pertaining to Mechanics and Local Motions* (1638). In the *Dialogue,* Galileo contrasted the accepted Ptolemaic model of the universe (the earth is the center of the universe) with the more recently developed Copernican model (The sun is the center of the universe). He strongly and very effectively supported the latter. In the *Discourses,* he attacked the analysis of the world as carried out by the Aristotelians. It was in this work that Galileo dealt with motion.

Although Galileo's works referred to physics and astronomy, his main goal was to introduce a new way (actually a rebirth of the Hellenistic approach exemplified by Archimedes, Aristarchus, etc.) of dealing with the world. This new way was based on observation and reason rather than appeal to authority and tradition. His writings were very important in the spread of this new way of coming to understand the world, for two reasons. He wrote in Italian rather than Latin, and he wrote in the form of readable dialogues—conversations rather than scholarly texts. Both of these contributed to a new phenomenon: Scholarly books became available to literate nonscholars who could read Italian but not Latin. Before Galileo's books, scholarly texts were readable only by those who could read Latin and who had the training to appreciate the complex arguments—in effect, only those who were associated with the Church or with the universities. Galileo was trying to reach an entirely different audience, literate people who were not part of the scholarly class.

Perhaps the following section will demonstrate how Galileo used the dialogue technique in advancing his philosophy. Three characters are involved in the *Discourses*: Salviati (representing Galileo), Simplicius (representing the Aristotelians), and Sagredo (a disinterested, intellectually curious bystander). Salviati and Simplicius argue, each trying to convince Sagredo that he is right and the other is wrong. In one of the episodes, Galileo attacks the concept of natural motion. The following exchange (paraphrased) is found:

SALVIATI: Please describe the motion of a moving box-shaped object that is free of external forces.

SIMPLICIUS: If the object is not supported on a surface, then it will fall straight down, because it is composed of Earth.

SAL.: Suppose that it is on a surface?

SIMP.: If the object is sliding down along the surface, it will continue to slide down, going faster and faster because it is composed of Earth and its natural motion is down. If the object were sliding up along the surface, it will gradually slow down, eventually come to a stop, and then slide back down as described before, again because of its natural motion down.

SAL.: Suppose that the surface is such that the object is neither sliding up nor down, but remains the same distance from the earth?

SIMP.: The object will slide, slow down, and gradually come to a stop because of friction between the surfaces.

SAL.:   Suppose that the surfaces are carefully polished?

SIMP.:   Then the object will slide farther before it slows down and stops.

SAL.:   And if the two surfaces are made as smooth as possible and perhaps some fine oil is placed on the surfaces?

SIMP.:   Then the object will slide even farther.

SAL.:   As the surfaces were polished, frictional forces were reduced. As the surfaces were polished even more and as they were lubricated with fine oil, friction was reduced to almost zero. Therefore the external agent has been reduced, and yet the motion continued. **Here is an example of motion that does not require an external agent and yet is clearly not natural motion, that is, it is not either up or down.**

Here, Galileo argued very persuasively that force does not cause motion. As we shall see later, he advanced the idea that force causes motion to change. Thus, if an object speeds up, slows down, or changes direction, its motion changes, and it must be that a force was acting on it to cause the change. If there is no force acting on an object or if all of the forces that are acting on it happen to cancel, the motion should not change. This idea was later explicitly expressed as **Newton's first law:**

If the total force acting on an object is zero, then that object will exhibit constant motion. This means that if it is at rest, it will remain at rest. If it is moving, it will continue to move at constant speed, in a straight line.

EXAMPLE 1.1

Consider a car that is driving on an icy road. Let us assume that, because of the ice, there is no friction. The driver may attempt to stop the car by applying the brakes. However, Newton's first law can be applied to explain what will happen. The force of gravity (the weight of the car) acting on the car will be balanced by the force of the road pushing up on the car. We have already noted that there is no friction. Therefore, there is no net force acting on the car so the motion of the car will remain constant. Even though the driver applies the brakes and hence stops the wheels from turning, the car will continue at the same speed, sliding along the road. If the driver were to turn the steering wheel, attempting to steer, the car would still continue on a straight-line path. It is important to understand that forces associated with the brakes of a car or a steering wheel represent *internal forces,* **forces within the car.** For the motion of the car to change, it is necessary that external forces play a role. If there are no external forces or if they cancel (the net force then being zero), the motion of the object will be constant. It will travel at a constant speed in a straight line.

Galileo also argued that motion itself was ambiguous. He maintained that motion is relative, not absolute. Whether something is moving or which way it is moving depends on the motion of the observer. (This concept was generalized almost 300 years later by Albert Einstein (1879–1955) in his special theory of relativity). Therefore, motion should not be taken as a basic quantity on which an understanding of the world would be based.

EXAMPLE 1.2

It is not always clear which way something is moving or even whether it is moving at all. When I was growing up, I lived on Long Island, across the river from New York City. The Long Island Railroad went through a tunnel under the river, and it was not unusual for a train to stop in the tunnel. Try to imagine sitting in a well-lit railroad car that is located in a very dark tunnel. If I looked out a window, all I saw was a reflection of the interior of the car. After a while, I looked out the window and saw people sitting in another train, also at rest, on the next track. After reading a newspaper for a while, I glanced out the window and saw that the other train was moving forward. I thought, "It's always the same, I get on the wrong train." But was the other train really moving forward? There were several possibilities:

1. My train was at rest, and the other train was moving forward.
2. Both trains were moving forward, but the other train was going faster.
3. The other train was at rest, and my train was going backward.
4. Both trains were moving backward, but my train was going backward faster.

This shows that motion, which is based on observation, is ambiguous and therefore cannot be taken as the basis of a rational understanding of the world. Galileo suggested that since motion is ambiguous, the basic quantity should be the change of motion. Thus instead of thinking that force causes motion, we shall accept as a basic principle that force is related to a change of motion. If an object's motion were observed to change, then there must have been a force acting on the object, that force causing the motion to change.

For example, as the earth moves around the sun, the direction of its motion is changing. This is caused by the force of gravity. As a car travels around a curve, the direction of its motion is changing. This is caused by the force of friction between the road and the tires. The force of friction plays a very large role in our world. In some situations, it is the force that makes a car slow down; in others, it is the force that makes the car speed up. If a person wants to be able to change motion quickly, as a football player does, the force of friction should be maximized, for example, by wearing cleats.

EXAMPLE 1.3

A person who sits up suddenly from a supine position may pass out. Why? The brain needs a constant supply of oxygen, carried by the blood. When a person sits up suddenly, the motion of the head (and the brain) has changed. It was at rest, and now it is moving upward. The force that causes this change of motion is supplied by the vertebrae that supports the head. The vertebrae push up on the head, causing it to start moving. The motion of the blood that is in the arteries, going to the brain, must similarly change to keep up with the head. The vertebrae cannot push on the blood; it is the heart that supplies the pressure required to push the blood up to the brain. If a person has low blood pressure or if, for some reason, the heart does not react quickly enough to the head's change of motion, then the blood will get left behind. The brain will not have enough oxygen, and unconsciousness may result.

Galileo, by showing that the neo-Aristotelians were wrong about motion, attacked the traditional way of gaining knowledge. To replace it, he advocated the idea that knowledge should be gained through observation and mathematical reasoning. As we have pointed out earlier, this was not a new idea. For example, Aristarchus had, in about 270 B.C., argued that the earth went around the sun. Eratosthenes (c. 273 B.C.–c.192 B.C.) determined, on the basis of measurements of lengths of shadows, that the circumference of the earth was 252,000 *stades* (approximately 24,000 miles). As with so much of the knowledge gained by the Hellenistic peoples, these ideas were "lost" with the fall of the Roman Empire, to resurface hundreds of years later during the Crusades and the centuries following.

To follow this development, we must make a transition from the qualitative discussion above to a quantitative analysis. To begin this process, we must establish more precise definitions of "motion" and "change in motion." These terms will be replaced by "velocity" and "acceleration."

Knowing the velocity of an object means that we know which way it is moving as well as how fast it is moving. Thus a car driving from Hartford to New Haven might have a velocity of 45 mi/h south, and the same car, later driving from New Haven to Hartford, might have a velocity of 45 mi/h north. In both cases, the car has the same speed (45 miles/hour) but different velocities, because of the different directions. Thus the motion of the car has changed. A person driving a car south who makes a U-turn to then drive north has effected a change in motion, an acceleration.

According to Galileo, there must be a force acting on the car that caused this acceleration. It would be the force of friction between the car's tires and the road.

When the velocity of an object changes (by speeding up, slowing down, or changing direction), we say that the object has *Accelerated*. This term, "acceleration," may well cause confusion among many people. This commonly happens when a word that was developed with a very narrow, specific definition passes into the realm of general conversation. A car has an accelerator. You use the accelerator to make the car go faster; if you take your foot off the accelerator, the car slows down. So we build a natural connection between acceleration and speed. More acceleration means more speed, and less acceleration means less speed. **Unfortunately this is not correct usage of the word.**

We will now use the driving of a car to show that acceleration and speed are not the same. Imagine two cars: One is a small three-cylinder car intended for use in city driving only; the other is a sports car, intended for open road driving. One of the many ways in which cars are compared is the time that is taken for the car to go from zero to 60 mi/h. Now suppose that you are driving onto a highway from an entrance ramp and must merge with the traffic. Some highways have entrance ramps that bring entering cars into the left lane (the high-speed lane) rather than the right lane. Suppose that you are stopped on the ramp, watching the traffic stream by at 65 mi/h, and you want to merge with it. You see a space between two cars and drive into it. Your car will have to be moving at the same speed as the traffic flow before the car behind you catches up. This means that the speed of your car will have to change a lot in very little time. Therefore your car will need a high acceleration. If we think of carrying out this merge with each of the two cars described above, it should be clear that *even though both cars reach the same speed,* you would be much safer in the sports car. It is capable of significantly higher acceleration than the other car and therefore is able to reach the required speed in a shorter time.

An operational way to realize that speed and acceleration are two different quantities is the reaction of your body. Your body does not react to speed. Imagine that you are sitting in an airplane that is flying in a straight line at a speed of 300 mi/h. This is certainly very fast, and yet, as you sit in the cabin of the plane, your body reacts as if you were not moving at all. If it were not for the noise and vibration caused by the engines, you might not even know that you were moving. If you pour some soda from a can into a cup, it flows exactly as if you were sitting at home at a table. There is no perception of motion, even though you are certainly moving at a very high speed. Or you could consider your body's reaction to the speed at which you are moving by virtue of the rotation of the Earth. Hartford, Connecticut, is located at 41.77° North latitude. As the earth rotates, Hartford moves around a circle of circumference 18,646 miles in 24 hours. Thus the city, and everything in it, is moving at 777 mi/h. Yet if you are sitting in a classroom in Hartford, there is absolutely no effect on your body. This speed, even though it is very high, is not perceptible

Now consider acceleration. When a plane speeds up, just before leaving the ground, you seem to be pushed back into the seat. If the airplane should suddenly change direction, you will be very aware of it. You may be thrown off your feet if you happen to be standing in the aisle, or you may feel yourself being pushed to one side or the other in your seat. If you happen to be pouring soda when the turn is made, the soda may miss the cup. A can of soda on the tray may slide off onto the floor. If the plane suddenly loses altitude, your stomach might seem to be coming up out of your throat. When the engines are used to slow the plane immediately after touchdown, you seem to feel a force pushing you forward. There is no question that **your body reacts to acceleration but not to speed.** This is just one way of realizing that speed (or motion) and acceleration are different quantities and the terms may not be used interchangeably. To deal with this area of potential confusion, we must be very precise in the use of words. **Acceleration means that the object's speed and/or direction is changing. Constant speed in a straight line implies zero acceleration.**

Galileo reasoned that an object's velocity cannot change instantaneously. That is, some interval of time, denoted by Δt, must pass in order for an object's velocity to change. **Acceleration is defined as the ratio of the change of the velocity to the time interval during which the velocity changed.** It is represented by the formula

$$\vec{a} = \frac{\Delta \vec{v}}{\Delta t}$$

(*NOTE:* The arrow over the symbols above identifies them as vectors. This special type of quantity, that includes both direction and magnitude, will be discussed in detail later. For now, we will just carry the arrows along.)

The symbol Δ is the capital Greek letter delta. It is the standard mathematical symbol for change and usually implies subtraction. **Note that the subtraction is always carried out in the following order: Later value − Earlier value.** For example,

$$\Delta \vec{V} = \vec{V}_2 - \vec{V}_1$$
$$\Delta t = t_2 - t_1$$

where the subscript 1 identifies an earlier situation (a pitcher releasing a ball) and the subscript 2 identifies a later situation (the ball reaching the catcher). $\Delta \vec{V}$ is the **change in the**

velocity of the object, and Δt is the change in time, or the time interval, during which the velocity changed. A negative Δt implies time going backward; however, that is a topic not covered in this book.

A particular acceleration, familiar to most of us, is due to the force of gravity between the Earth and any object near it. For example, if a rock is released from rest, it will fall, and its speed will increase; thus it is accelerating. Many measurements have been made of the acceleration due to gravity as exhibited by a large number of objects. Surprisingly, it turns out that (under ideal circumstances) all objects near the Earth experience the same value of the acceleration of gravity. This means that a Ping-Pong ball, a golf ball, and a spherical piece of lead of the same diameter will all have the same acceleration due to gravity even though they contain quite different amounts of matter. This violates common sense; however, careful measurements show that it is true.

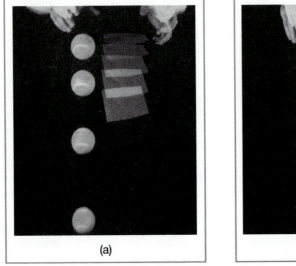

(a)

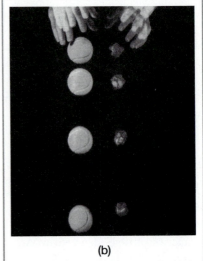

(b)

The figures show what happens when a small ball and a piece of paper are dropped. Each figure shows successive snapshots (approximately 0.1 s apart). The figure on the left shows that the ball falls through increasingly large distances in each successive 0.1 s. It falls farther in the second 0.1 s than in the first 0.1 s and so forth. This shows that the ball is moving faster during the second 0.1 s than during the first 0.1 s, and so forth. Thus the ball is accelerating.

Careful measurements show that although the ball's speed is increasing, the amount of the increase in each of the time intervals is the same. This leads to the idea that an increasing velocity can be associated with a constant acceleration. The piece of paper (again referring to the figure on the left) does not seem to be moving as much as the ball. Although it starts out with the ball, it quickly gets left behind. It is still moving, but it is moving more slowly than the ball. It seems that although the paper's velocity is increasing, its increase is less than that of the ball.

Now look at the picture on the right. We have the same ball and the same piece of paper. But the paper has now been crushed into a spherical clump before it is dropped. The figure shows that the ball and the clumped piece of paper keep in step. Each of them has the same increase of velocity. Comparison of the two figures shows the effect of air resistance. The piece of paper experiences a lot of air resistance when it is flat. However, when the paper is clumped into a ball, the effect of air resistance decreases noticeably. These figures represent one of Galileo's radically new ideas: If air resistance could be removed, all falling objects would experience the same change of velocity, that is, the same acceleration. This particular acceleration is called the acceleration of gravity.

The **acceleration of gravity** has attracted considerable experimental and theoretical investigation. It has been assigned a special symbol, $\vec{g}$, and its magnitude has been generally accepted to be 9.8 m/s^2 (meters per second squared), or equivalently, 32 ft/s^2, at the Earth's surface, and is always directed down (toward the center of the Earth).

EXAMPLE 1.4

Imagine someone throwing a ball straight up into the air. Intuitively, we know that the ball will slow down and come to stop when it reaches its maximum height. The ball then comes back down to the thrower, moving faster as it falls. We will now use the three formulas above and see how the calculated results compare with our intuition. Assume that the initial velocity of the ball is 40 ft/s, *up*. As was stated earlier, the acceleration of gravity will be 32 ft/s^2, *down*.

1. Rewrite the given information with mathematical symbols:

$$\vec{v}_1 = 40\,\frac{\text{ft}}{\text{s}},\ \text{up}$$

$$\vec{v}_2 = ?$$

$$t_1 = 0$$

$$\vec{a} = 32\,\frac{\text{ft}}{\text{s}^2},\ \text{down}$$

$$t_2 = ?$$

2. Basic equations:

$$\vec{a} = \frac{\Delta\vec{v}}{\Delta t}$$

$$\vec{a} = \frac{\vec{v}_2 - \vec{v}_1}{t_2 - t_1}$$

3. Substitution:

$$32\,\frac{\text{ft}}{\text{s}^2},\ \text{down} = \frac{\vec{v}_2 - \left(40\,\frac{\text{ft}}{\text{s}},\ \text{up}\right)}{t_2 - 0}$$

To avoid carrying the arrows and the "up" and "down" along, we will use a standard convention: Designate the magnitude of any vector that points upward as a positive number and the magnitude of any vector that points downward as a negative number. To reduce the amount of writing, we will also drop the units.

$$\uparrow+ \quad (-32) = \frac{\vec{v}_2 - (40)}{t_2}$$

This equation may be solved for v_2:

$$v_2 = 40 - 32t_2$$

We can now substitute values for t_2 and calculate corresponding values for v_2:

t_2 (s)	v_2 (ft/s)
0.0	40
0.25	32
0.5	24
1.0	8
1.5	−8
1.75	−16
2.0	−24
2.25	−32
2.5	−40

The results of the calculations, as shown in the list, are consistent with the predictions that were based on intuition. Recall that + values of velocity mean that the ball is moving up and − values mean that the ball is moving down. Notice that at 1.0 s, the ball is moving up, and at 1.5 s, the ball is moving down. At some point between 1.0 s and 1.5 s, it must have stopped. We can determine when it stopped by setting v_2 equal to 0 and solving for t_2:

$$v_2 = 40 - 32t_2$$
$$0 = 40 - 32t_2$$
$$t_2 = 1.25 \text{ s}$$

Thus we now know that the ball reaches its maximum height 1.25 s after it was thrown. Notice that at 2.5 s, the ball is moving at 40 ft/s, down. This implies that, after falling for the same amount of time that it spent rising, the ball is moving down as fast as it was originally thrown upward. This is not correct, and it is very important that we discuss this point.

EXAMPLE 1.5

Imagine a batter hitting a fly ball to the outfield in a baseball game. The ball travels to an outfielder, who casually catches it at the same level as the one from which it was hit. The ball has returned to its original level, and on the basis of Example 1.4, we might expect that it is moving at its initial speed. This cannot be true. Although an outfielder may catch the ball quite casually, things would be quite different if the pitcher were to try to stop the ball immediately after it had been hit. The ball is moving much faster when it is hit than when it reaches the outfielder. So what was wrong with our calculation in Example 1.4?

We assumed that the only force acting was the force of gravity and therefore did not consider friction between the ball and the air (air resistance). This force certainly exists and will slow the ball. This explains why it is moving much more slowly when it gets to the outfield than when it passes the pitcher. So why leave it out of the calculation?

We are faced with two choices. One possibility would be to represent the situation as completely as possible; for example, include air resistance (and face the possibility that we may not be able to solve the resulting equations). The other possibility is to represent the situation with some simplifying assumptions; for example, ignore air resistance (thus resulting in solvable equations). Quite often, particularly in dealing with something as complicated as the human body, it is necessary to make the second choice. This is an example of a process (called idealizing a sit-

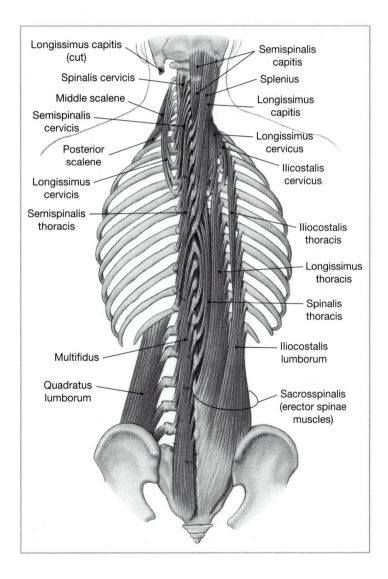

uation) that is employed whenever a real-world situation is represented by mathematics (called mathematical modeling).

For example, the figure on page 11 shows the major muscles that act within a person's back. Clearly, there are several muscles that originate on the hip and connect to many different points on the skeleton of the trunk. It would be extremely difficult to use all of these muscles in an analysis of extension of the back. As we shall see later (see page 107), a commonly made assumption is that there is only one muscle (represented by the *erector spinae*) playing a role in this extension. This is not strictly true, but it leads to equations that are solvable, and the results agree fairly closely with measurements.

The issue of using a model rather than a maximally complete representation of a real-world problem will arise many times in this book. It is very common when one attempts to use mathematical analysis while dealing with the human body. The prime justification for using approximations, neglecting forces, and so forth is that the results agree reasonably well with measurements. For more detailed analyses, a simple model (at the level used in this text) may not suffice. More involved models, and hence more involved mathematical analysis, must then be used.

Acceleration in g's

Quite commonly, an acceleration—for example, the acceleration of an airplane or an amusement park ride—is expressed as a multiple of the acceleration of gravity. When this is done, the acceleration is referred to as so many **g's**:

$$\vec{\mathbf{a}}_{g's} = \frac{\vec{\mathbf{a}}}{g}$$

This has three advantages:

1. An unknown acceleration is referred to in terms of a familiar acceleration.
2. Since we are dealing with a ratio, the units will cancel.
3. As we shall see later, the net force acting on an object is easily calculated as the weight of the object multiplied by its acceleration in g's:

$$\vec{\mathbf{F}} = \vec{\mathbf{a}}_{g's} W$$

EXAMPLE 1.6

During the 1920s, there was particularly intense competition among amusement parks. In particular, each of them wanted to have the most terrifying roller coaster. Competing rides were built higher, steeper, and with more sharp turns as the decade passed. There was no way to determine just how steep or how sharp a turn was safe, and so designers depended on the response of the consumers. It was determined, after many injuries and lawsuits, that the maximum acceptable acceleration for a person riding in an amusement park ride was 4.2 g's. How much net force would be exerted on a 120-pound rider in such a ride?

$$F = (a_{g's})(\text{weight})$$
$$F = (4.2)(120 \text{ lb})$$
$$F = 504 \text{ lb}$$

This force was usually exerted by the safety bar that kept the passenger within the car.

Units

As we have seen above, the value of the acceleration of gravity depends on the units used, whether m/s² or ft/s². Unfortunately, several systems of units are used, and we have to deal with this situation. The two systems that we shall use in our discussions are the everyday system used in the United States and the system (called the metric system, or the *Systeme Internationale*, or SI, system) used in most of the rest of the world. Both systems will be used in this book. This usage may lead to some confusion, but it is necessary because it represents the way that most students who are using the book think about units. I am sure that all of you are very comfortable with the idea of pounds. You realize that lifting a 5-pound weight is very reasonable while lifting a 500-pound weight may be impossible. However, you might not realize that lifting a 22-newton weight (equivalent to the 5 pounds) is very reasonable. Similarly a speed of 45 miles per hour is probably more understandable than a speed of 20 meters per second (even though they represent the same speed). However, the SI system, the legal system in the United States and the most commonly used system in the world, cannot be ignored. Therefore, you must develop the facility to work with both systems.

Only three basic units are needed for most of the work we shall do in this course. We will need a unit of length, a unit of time, and a unit of "quantity of stuff." (We shall later introduce a fourth quantity: temperature.)

UNIT	USA	SI
length (or distance)	foot	meter
time	second	second
quantity	pound	kilogram

The unit of quantity will be a source of trouble. The USA unit, the pound (lb), is a measure of **weight** while the SI unit, the kilogram (kg), is a unit of **mass. Weight and mass are not the same.** They are related by the equation

$$W = mg$$

where g is the acceleration of gravity, described above. The pound is associated with W, and the kilogram is associated with m. If we use the pound, then we must use 32 ft/s² for g, but if we use the kilogram, then we must use 9.8 m/s² for g. This is messy, and it would certainly be simpler if everyone used the same system, but that is not what happens. So if a person weighs ("weight" is a tip-off word that means "force") 150 lb, we could ask, "What is the person's mass?" Alternatively, if a person has a mass of 70 kg, we could ask, "What is the person's weight?" The tricky thing is that while the mass of an object is constant, that is, it has the same value everywhere, the weight of an object is variable; that is, the weight has different values, depending on where the object is. For example, you would weigh less at the top of Mount Washington than you would weigh at the bottom of the mountain. (This will be explained later when we discuss the force of gravity). But your mass would be the same at both places. This will be a major source of difficulty, and I will repeat it for emphasis:

Weight and mass are not the same:

- The mass of an object is constant.
- The weight has different values, depending on the location of the object.
- Weight and mass are related by the equation

$$W = mg$$

EXAMPLE 1.7

A very spectacular example of the difference between weight and mass occurred during a Space Shuttle mission in 1992 when one of the major goals was to capture a communications satellite that had not gone into the proper orbit. The idea was to capture the satellite, bring it into the cargo bay, fit it with a rocket motor, place it back out into space, and then fire the rocket that would push the satellite into the proper orbit. The effort to use the control arm on the Shuttle to capture the satellite failed, and it was necessary for three of the astronauts to go out into space and pull the satellite into the cargo bay by hand. It would not seem reasonable for three people to be strong enough to carry out such a task because the satellite weighed 4.5 tons on the Earth. However, in orbit, the satellite was "weightless". As far as the astronauts were concerned, the satellite had little or no weight, but it had a tremendous amount of mass.

The difference is that while weight is a measure of the force of gravity (to be discussed in detail later) on an object, mass is a measure of the amount of material in the object. Weight changes with the object's location and motion, but mass does not change. These two terms are not interchangeable and must not be confused.

We may now extend the table given earlier

UNIT	USA SYSTEM	SI SYSTEM
length (or distance)	foot (ft)	meter (m)
time	second (s)	second (s)
weight (force)	pound (lb)	Newton (N)
mass	slug (sl)	kilogram (kg)
g	$32 \dfrac{ft}{s^2}$	$9.8 \dfrac{m}{s^2}$

Note the new units: slug (mass) and Newton (force).

The importance of familiarity with these two systems is illustrated by an unfortunate happening that was reported in the press on or about October 1, 1999. It had taken four years and $125 million to build the Mars Climate Orbiter. All went well with its launch and its trip as the orbiter approached Mars. It then began its first orbit. It passed behind Mars and never reappeared. A very intense investigation determined that the failure of the orbiter was due to confusion regarding units. The spaceship had been designed and built by the Lockheed Martin Company but, while in space, was controlled by NASA's Jet

Propulsion Laboratory. **Unfortunately, the data regarding the required thrust of the rockets had been specified in pounds by the designers but was interpreted as Newtons by the controllers.** So when the orbiter fired its rockets to establish a safe orbit, the rockets produced too much thrust, and the orbiter probably broke up or burnt up in Mars's atmosphere.

Unit Conversion

In dealing with real-world examples, we shall have to convert units from one type to another (see the list in Appendix 1).

EXAMPLE 1.8

"I weigh 150 pounds. What is my weight in Newtons?" There are many techniques that can be used to change units. One of the easiest is to use a calculator, such as a TI-85, that has this functionality built in. If you do not have access to such a calculator, I suggest that you use the following technique:

Step 1. Write the quantity (with units) that you wish to change within parentheses.

$$(150 \text{ lb})$$

Step 2. Multiply this quantity by a term in parentheses that contains a ratio, the denominator being the unit that you want to get rid of and the numerator being the unit with which you want to replace it.

$$(150 \text{ lb}) \left(\frac{?\_N}{?\_\text{lb}} \right)$$

Step 3. The term in the second set of parentheses must be equivalent to 1 so that you do not change the magnitude of the original quantity (you are only changing the units). **This can be assured if the numerator is equal to the denominator.** Use the data given in Appendix 1 to determine how many Newtons correspond to how many pounds. From the chart, we see that 1 N is equal to 0.225 lb. Therefore, the numerator should be 1 N, and the denominator should be 0.225 lb:

$$(150 \text{ lb}) \left(\frac{1 \text{ N}}{0.225 \text{ lb}} \right)$$

Step 4. Notice that the "lb" will cancel, and the result will be 667 N. You can see why people would rather express their weight in pounds than in Newtons.

Note that a McDonald's Quarterpounder® might reasonably be renamed the Newtonburger.

EXAMPLE 1.9

A household is billed for 350 kilowatt-hours (kWh) of electrical energy during a 30-day month. Determine the average rate at which energy was used, in horsepower.

Solution

From the list in Appendix 1, we see that

$$0.738 \text{ lb ft} = 2.78 \times 10^{-7} \text{ kWh}$$

$$0.738 \text{ lb ft/s} = 0.00134 \text{ hp}$$

$$(350 \text{ kWh})\left(\frac{0.738 \text{ lb ft}}{2.78 \times 10^{-7} \text{ kWh}}\right)\left(\frac{0.00134 \text{ hp}}{0.738 \frac{\text{lb ft}}{\text{s}}}\right) = 1{,}687{,}050 \text{ hp s}$$

We are close; we have an answer that has units of horsepower times seconds. To get rid of the seconds, we must divide by a time interval. There is one given in the problem: 30 days.

$$\left(\frac{(1{,}687{,}050 \text{ hp s})}{(30 \text{ days})}\right)\left(\frac{1 \text{ day}}{24 \text{ h}}\right)\left(\frac{1 \text{ h}}{3600 \text{ s}}\right) = 0.65 \text{ hp}$$

Thus, the household is using energy at the average rate of 0.65 horsepower.

To summarize: The suggested procedure to be used to convert units is to multiply the quantity by one or more ratios, each being equivalent to 1, such as

$$\left(\frac{1000 \text{ liters}}{35.31 \text{ ft}^3}\right) \quad \text{or} \quad \frac{0.000948 \text{ BTU}}{0.239 \text{ cal}}$$

The object of using these ratios is to arrange them such that the unwanted units cancel and the desired units remain.

VERY IMPORTANT NOTE

In beginning any quantitative analysis, one of the early steps in the analysis should be to deal with units. I recommend that you pick either one of the two systems described above. Having chosen a system, write all of the given information in that system of units, carrying out conversions where necessary. The result of the analysis will then automatically work out with units in that system.

We have discussed acceleration and its relation to velocity. However, there are relations and definitions that are necessary for the mathematical description of motion. These new terms are *displacement* and *average velocity*. They will be discussed in detail in the section on vectors (pages 27–50).

FORCE

Basics of Force

What is a force?

A force is a push or a pull. In classical physics, it is described as an interaction between two objects. So the force of gravity is the interaction between us and the earth or between the earth and the moon. The force of friction is an interaction between the soles of my shoes and the floor. The force of my biceps muscle is an interaction between the muscle and my forearm.

What does a force do?

A force acting on an object can cause the motion (velocity) of the object to change. That is, it can make the object speed up, slow down, or change direction. Thus the force of gravity between the earth and a rock can make the rock speed up (if it is moving down), slow down (if it is moving up), or change direction (if it is moving horizontally).

A force can also cause an object to change shape. For example, the weight of a standing person can cause the length of the person's spinal column to decrease; conversely, astronauts actually grow a little taller while spending extended time far from the earth.

If there are several forces acting on an object, they operate together and may reinforce each other or perhaps cancel each other. In our bodies, we have many forces acting at the same place at the same time. For example, consider the forearm. The *biceps* muscle may be pulling it up, gravity is always pulling it down and a rock, held in the hand, would be pushing it down. In many of the problems that we shall consider, forces such as these will cancel (thus summing to zero).

Where do forces come from?

Although our concept of force comes from observing what goes on in the world, our explanation of force is theoretical; it comes from imagination. Physicists picture observable forces as arising from combinations of a very small number of fundamental forces. These fundamental forces are gravity, electricity, magnetism, the weak nuclear force, and the strong nuclear force. The last two of these are used only in trying to analyze the details of a nuclear reaction, such as radioactive decay. Albert Einstein showed, as part of his Special Theory of Relativity (1905), that electricity and magnetism are two different labels for the same force. It is this force that represents the interactions between atoms and molecules that leads eventually to the structure and function of virtually all of the objects (those larger than the nucleus of an atom) in the world. Among these are the components of the muscular-skeletal system in your body. When we speak of the strength of a muscle or of a bone, it is really the electric force that we are describing. Although the forces exerted by bone and by soft tissue (muscle, tendon, etc.) are basically the same (both being electric in nature), they appear quite different. As we shall see later, the force exerted by bone is always compression: It pushes. The force exerted by soft tissue is tension: It pulls.

It is particularly interesting that the most important force exerted on the body by the outside world, the force of gravity, was not even recognized as a force until the end of the seventeenth century.

Force of Gravity

As we have already seen (pages 3–9), Galileo maintained that any object exhibiting an acceleration must be acted upon by an external force. He also taught that, in the absence of air, all objects falling (free fall) near the surface of the earth would experience the same acceleration. He did not, however, identify the force that caused this acceleration. At about the same time that Galileo was carrying out his research, a German mathematician named Johannes Kepler (1571–1630) was trying to explain the apparent motions of the planets.

It was then generally believed that the earth was the center of the universe and that the sun, the moon, and all of the planets moved in circular paths around the earth. As more and more astronomical data were gathered during the centuries leading up to the sixteenth century, it became more and more difficult to accept this commonsense model. In 1543, a Polish mathematician, Nicolaus Copernicus (1473–1543) published a book, *De Revolutionibus Orbium Caelestium (On the Revolutions of the Heavenly Spheres,* often referred to as *Revolutionibus)* setting forth the idea that the sun was the center of the universe and that the earth and the other planets went around the sun in circular orbits. By coincidence, this was the same year in which Vesalius published his major work (page xi).

In trying to justify Copernicus's model, Kepler, using the best data then available, arrived at some startling conclusions. He found that the data could best be explained by making three assumptions:

1. The planets traveled around the sun. They moved in elliptical orbits with the sun at one focus. This was a particularly heroic statement, as there had been a very long tradition associating circles with the perfection of the heavens. Ellipses carried no such imagery.

2. The planets moved faster in their orbits when they are closer to the sun than when farther from the sun. In particular, each of the two arcs (*AB* and *CD*) represents one month's motion of the planet in its orbit. According to this law, the two wedges *A-B*-sun and *C-D*-sun have equal areas. Clearly, the radius of the orbit of each planet swept out equal areas in equal time intervals. In the figure, the planet is moving faster during the *CD* month than during the *AB* month.

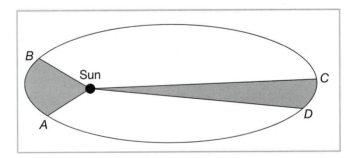

3. The orbital radii and orbital periods of the planets were related to each other. Kepler found a mathematical relation between the average radius (*R*) of each orbit and the time (*T*) that it took for that planet to complete one orbit:

$$R \propto T^{\frac{2}{3}}$$

These results have come to be known as Kepler's three laws. They suggested that there was a high degree of order in both the distribution and the motion of the planets. This both satisfied Kepler's religious beliefs and encouraged him to find an explanation for this order. He suggested that there was some sort of emanation from the sun causing the planets to follow the observed orbits. This emanation was not light but something else by means of which the sun influenced the planets. The idea that, the sun somehow affected the planets and caused their orbits was the basis of Newton's invention of the force of gravity.

To summarize, at this point in history, there were several observations or ideas that seemed to have no theoretical basis. In fact, they were not understood at all; they lay outside of the accepted world view. These included the idea (due to Galileo) that objects of different masses could have the same free fall acceleration and Kepler's three laws. All of these ideas were supported by observations and data. However, they made no sense at the time.

These problems, as well as several others, were cleared up in 1687 when Newton published the *Principia*. In this book, he set forth an amazing idea: There was a force, the force of gravity, acting between every pair of objects in the universe. This force was attractive, pulling the objects together. Newton showed mathematically that Galileo's ideas and Kepler's laws could be explained if the force were proportional to the product of the two masses and inversely proportional to the square of their separation. Expressed using standard symbols this becomes

$$F = \frac{GM_1 M_2}{r^2}$$

where G is a universal constant, the M's are the masses of the two objects, and r is the distance between them. G could not be experimentally determined in Newton's time, but he was able to estimate its value closely enough to establish the validity of the formula.

In 1705, his friend Edmond Halley used Newton's method of analysis and the force of gravity to show that what had been thought to be sightings of several comets (1531, 1607, and 1682) were actually different sightings of the same comet. He used the mathematics to predict that the comet would be sighted again in 1758. It was observed on Christmas night, 1758, and was named Halley's Comet in his honor. This amazing prediction was accepted as proof that Newton had been right and that there really is a universal force of gravity represented by his equation.

In 1798, another Englishman, Henry Cavendish, carried out a series of experiments that were later shown to yield a numerical value of G. The experiments have been repeated many times, and the currently accepted value of G is 6.673×10^{-11} in the SI system.

Thus, according to Newton's law of gravitation, the earth is pulling down on each part of your body, and you are pulling up on the earth. Of course, the earth is much more massive than your body and so is not measurably affected by this force. Your body, on the other hand, is greatly affected. The constant downward force has major effects on your heart, muscles, and bones.

One of the most important functions of the heart is to pump blood up to the brain. Since we spend most of our time erect, the brain is above the heart. Most other animals' brains are typically on the same level as the heart. Since gravity is constantly pulling blood downward, the heart must force the blood up to the brain. This puts a great stress on the heart, and when the process fails, the result is loss of consciousness and sometimes death.

A very interesting and potentially very important example of the interaction of gravity and blood flow is the circulation of blood in a giraffe. Giraffes are the tallest land animals;

their heads may be as much as 18 ft above the ground. The heart is located approximately halfway between the head and the feet—thus about 9 ft above the ground. This arrangement presents several questions about the giraffe:

1. How much pressure is required for the heart to pump blood up to the giraffe's brain?

2. How do the walls of the giraffe's carotid artery (the one that carries blood from the heart to the brain) deal with this high pressure?

3. Given that the giraffe's heart is about 9 ft above its feet, what prevents blood from pooling in its feet and the veins located within the legs?

4. When a giraffe drinks, it usually lowers its head to the water, spreading its front legs to allow the neck and head to lower. If it is startled, it immediately raises its head and gallops off. How does the circulatory system react so that the giraffe does not lose consciousness because insufficient oxygenated blood is getting to the brain?

It is now known that giraffes have much higher blood pressure than other animals; 260/160 for a giraffe compared to 120/80 for man. It is also known that the pressure in the tissues that surround the blood vessels is much higher in giraffes than in other animals. This higher extra-arterial pressure helps to protect the walls of the artery from the high blood pressure required to adequately supply the brain. Results of this type of study may enable us to deal better with high blood pressure and circulatory problems in people.

Newton's formula for the force of gravity, given above, shows that the force of gravity between two objects depends on the distance between them. If the distance is extremely large, perhaps infinite, the force of gravity would be zero. If each of the objects is a point mass, an object so small that its size is negligible, then the distance between them is unambiguous. However, if one of them is large, such as the earth, then the value of the distance between them is not obvious. Is it the distance between their nearest surfaces, the centers, or what? Newton was able to show, via his invention of integral calculus, that there is a location within any object at which, for the purpose of calculating a force of gravity, all of its mass may be assumed to be located. This point is called the **center of gravity** of the object. So the variable r in the formula is the distance between the two centers of gravity. If the mass of the object is uniformly distributed and if its shape is regular, such as a sphere, cylinder, or cube, the center of gravity is located at the geometric center. We shall use the concept of center of gravity a great deal, later in the course.

Since the earth is a relatively smooth sphere, the distance between someone and the center of the earth is pretty much the same, no matter where she is. Therefore, the force of gravity on her is effectively constant and is given a special name: her *weight*. This definition implies that zero weight implies zero force of gravity due to the earth. That, in turn, implies an infinite separation from the center of the earth.

The term "weight" can be easily misused. First of all, as we have already pointed out, it is commonly confused with "mass." A more up-to-date misuse is related to the space program. It is not unusual to read that astronauts are "weightless" when in orbit or that an experiment was carried out in a "microgravity" or—worse—"zero gravity" environment.

Let's compare the force of gravity on a person who is in orbit (assume 250 miles [402×10^3 m] above the surface of the earth) to the force of gravity on the same person when on the surface of the earth (radius = 4×10^3 miles = 6.44×10^6 m):

$$F_{\text{person on surface of earth}} = \frac{GM_{\text{earth}}M_{\text{person}}}{r_{\text{earth}}^2}$$

$$F_{\text{person 250 miles above earth}} = \frac{GM_{\text{earth}}M_{\text{person}}}{(r_{\text{earth}} + 250 \text{ miles})^2}$$

Taking the ratio of these two expressions, we get

$$\frac{F_{\text{person in orbit}}}{F_{\text{person on surface}}} = \frac{(r_{\text{earth}})^2}{(r_{\text{earth}} + 250 \text{ miles})^2}$$

$$\frac{F_{\text{person in orbit}}}{F_{\text{person on surface}}} = 0.89$$

This shows that the force of gravity from the earth on a person in orbit is about 89% of its value when the person is on the earth. A 150-lb person (on the earth) would therefore "weigh" 133 lb in orbit—hardly weightless.

A way out of this confusion is to define weight as the force that would be measured, for example, by a bathroom scale. It is easy to demonstrate that your "weight" as measured by a scale can vary. Take a scale into an Up elevator and read it before the elevator starts, as it is starting to move, as it comes to a stop, and after it has stopped. You will see that you seem to weigh more as the elevator is gaining speed and less as the elevator comes to a stop. Your weight will be "normal" when the elevator is at rest or moving at constant speed. If you try a Down elevator, your weight will appear to be less as the elevator is gaining speed and more as the elevator slows. These reading are consistent with your body's reaction. You feel "sort of weightless" as the elevator accelerates down but "heavier" as the elevator accelerates up. We shall see later, in the discussion of centripetal acceleration, that when someone is in orbit, her acceleration is "down," toward the center of the earth, and therefore she feels weightless, as in the elevator accelerating downward.

This conflict could be resolved by defining the weight of an object as the magnitude of the force of gravity that is exerted by the nearest planet and the apparent weight of an object as the weight that would be indicated by an appropriate measuring device such as a bathroom scale. However, this is too cumbersome for common use, and so one must be very careful about how the word "weight" is used.

One of the surprising results of the space program was the observation that when astronauts and cosmonauts returned from extended periods of time in orbit, they were very weak, sometimes needing help to walk. Medical investigation showed that they had lost bone mass and muscle mass. We now understand that as the body reacts to the constant pull of the force of gravity, it responds by building bone and muscle to resist this pull. When a person is in an environment in which either the force of gravity is very small (weightless) or the body is in free fall (apparently weightless), as occurs in orbit, the body does not build bone and muscle, and these tissues actually atrophy.

From the very basic point of view of forces as mentioned earlier, the force in a muscle and the force of friction are basically the same. They arise from the electrical force that holds one atom to another and to another until a piece of material, such as a muscle, a floor tile, or a shoe sole, is built up. The explanation of the various forces in the world in terms of these five basic forces requires rather advanced physics knowledge and is not necessary for our discussions. We shall concentrate not on where the forces come from, but on their effect and on how we represent this effect mathematically.

Friction

Whenever two surfaces are in contact, they interact with each other. The interaction is made up of mechanical and/or electric forces between the molecules and atoms that make up each surface. For example, if the two surfaces are very rough, we can imagine that the "hills" of one surface could fit into the "valleys" of the other surface. There would then be

a great deal of mechanical force preventing the two surfaces from sliding past each other. If the surfaces were very smooth, such mechanical forces could be minimized. If a material, such as gas or a nonviscous fluid separated the surfaces, there would be significantly less friction between them. This is how lubrication—for example, the oil in an automobile motor or the synovial fluid in a body's joints—works.

(*Note*: If the surfaces are so smooth that there is no room for even a layer of air between them, they will experience so much friction that they will appear to be welded together.)

Analyzing such a large number of forces is not realistic and is avoided by dealing with the cumulative effect of all of the interaction forces. This cumulative effect is called the force of friction. Our understanding of friction is based on the results of many experiments rather than on considerations of basic theory. On the basis of such experiments, we can make the following general statements:

1.   Friction may be divided into two fundamental categories depending on whether the surfaces are moving past each other (kinetic or sliding friction) or are not moving (static friction).

2. The magnitude of a force of friction may be calculated from the following equation:

$$F_f = \mu F_N$$

where F_f is the force of friction, μ is the coefficient of friction (to be determined from tables or experimentally), and F_N is the amount of force (always perpendicular to the interface between the two surfaces) that is pushing the two surfaces together. This force is referred to as the NORMAL FORCE. The normal force (F_N) has three characteristics:
   a. It is always directed perpendicular to the surface.
   b. It is directed toward the object.
   c. In general, it is not equal to the weight of the object. The magnitude of F_N must be determined in each situation.

3. The direction of the force of friction on a specific surface is given by the following rules:
   a. If the object is sliding, then the force of friction (called sliding or kinetic friction) on the object is directed opposite to the direction of the motion and its magnitude is given by

$$F_K = \mu_K F_N$$

   b. If the object is not sliding, then the force of friction (called static friction) on the object is directed opposite to the direction in which the object would slide if there were no friction. The force of static friction does not have a specific value, as does the force of kinetic friction. Rather, it is characterized by a maximum value.

$$F_S \leq \mu_S F_N$$

If this maximum value is exceeded, then the object will slide; static friction will no longer apply, as we will be dealing with kinetic friction.

The idea that the force of static friction is not constant but does have a maximum value is very important in trying to understand many forms of human activity, such as sports. In many situations in sports, it is necessary to speed up, come to a stop or change direction in as small a time interval as possible. For example, after hitting the ball, the batter cannot take a long time to reach top speed in running to first base; when evading a tackler, a running back cannot take a long time to change direction. Since the time interval should be as short as possible, the corresponding acceleration will be as large as possible. As we shall see later, this means that the external force that is causing the acceleration of the body will be as large as possible. The athlete learns from experience that to change speed quickly (i.e., in a short time interval) or to change direction quickly, it is necessary to make one or both of the feet push as hard as possible against the ground in the proper direction. The ground will then push back on the foot in the opposite direction (see the discussion of Newton's third law later in the chapter). It is the force that the ground exerts on the foot that causes the required acceleration. Since slipping must be avoided, it is static friction that represents the required force. Since this force has a maximum value (that depends on the nature of the two surfaces, that is, the sole of the shoe and the ground), athletes learn that if they push too hard on the ground, the result will not be the expected acceleration but, rather, slipping. Thus we can explain the use of cleats, spikes, or specially constructed soles of basketball shoes: If the surface of the sole of the shoe is properly modified, the maximum force of static friction will be larger, and thus a greater acceleration may be achieved.

COEFFICIENTS OF FRICTION	μ_s	μ_k
dry bone on bone		0.3 (Davidovits, 1975, p. 23)
bone on joint, lubricated		0.003 (Davidovits, 1975, p. 23)
steel on ice	0.02	0.01 (Davidovits, 1975, p. 23)
rubber on dry concrete	0.9	0.7 (Gustafson, 1980, p. 126)
rubber on wet concrete	0.6	0.4 (Gustafson, 1980, p. 126)
Teflon on steel	0.04	0.04 (Gustafson, 1980, p. 126)
bone with synovial fluid	0.016	0.015 (Gustafson, 1980, p. 126)
steel on steel (dry)	0.2	0.1 (Gustafson, 1980, p. 126)
steel on steel (oiled)	0.04	0.03 (Gustafson, 1980, p. 126)

EXAMPLE 1.10

If you slide your feet on the floor as you walk forward, there will be a force of kinetic friction between the surface of the floor and the bottom of your shoe. Since the shoe is sliding forward, the direction of the force of friction on the shoe is backward. This may also be discussed by the application of a fundamental principle of physics that is known as Newton's third law. This may be stated as, follows:

When two objects interact, they exert forces on each other. These forces are equal in magnitude but opposite in direction.

Thus since the bottom of the shoe is pushing the surface of the floor forward, the force on the bottom of the shoe will be backward. Therefore the force of friction of the floor acting on the shoe will have the effect of slowing the slide.

EXAMPLE 1.11

Consider the tires of a car. If the car is moving forward, the tires are rotating clockwise. The bottom surfaces of the tires do not usually slip on the surface of the road. This is why there is such a clear imprint of the tire tread on a dry roadway after a tire has moved through a puddle. Since the tire is not slipping, there will be a force of static friction between the road surface and the tire. If the car is speeding up, the direction of the force of static friction on the tire is forward. This is not intuitive but may be understood from the following discussion. If the tire happened to be on ice, the tire would spin, the bottom surface slipping backward. Thus, since in the absence of friction the tire would slip backward, the force of static friction on the tire must point forward. Or, by applying Newton's third law, we see that the tire is pushing the road surface backward, so the road surface is pushing the tire forward.

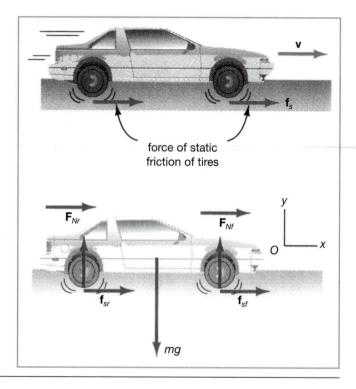

force of static
friction of tires

EXAMPLE **1.12**

There is little friction between the bones that make up the joints of the body. This is because they are separated by material, such as synovial fluid, that leads to reduced friction. As we saw above, when the force pushing the surfaces together increases, the result will be more friction. This is not the case in dealing with skeletal joints. When the force compressing a joint increases, a larger amount of synovial fluid is squeezed out of the cartilage that normally holds it, making it available to lubricate the joint. This results in a decrease in friction.

EXAMPLE **1.13**

Imagine that you are walking on a slippery surface, such as an icy sidewalk. You automatically take much smaller steps than you would if you were walking on a dry sidewalk. Why? Consider the trailing foot. It is behind you and is pushing down and backward on the sidewalk. The sidewalk reacts by pushing up and forward on your foot. The upward force is caused by the strength of the concrete to withstand compression. The forward force is caused by friction between the sidewalk and your foot. If the coefficient of friction is very small, as it would be for an icy surface, this forward force will be small, and your foot will slip backward. As we shall see later, when we deal with components of a vector, the amount of force that your foot exerts backward depends on the angle between your leg and the vertical. The larger the angle (i.e., bigger strides), the greater the horizontal force. A larger horizontal force exerted by your foot will require a larger force of friction to prevent slipping. Without knowing the physics behind the situation, you automatically take smaller steps, keeping the angle small, thus exerting less horizontal force and requiring less friction to prevent slipping.

PROBLEM SET 1

Note: Use the conversion factors given in Appendix 1 for changing units.

1.1. Describe the motion of an electron that is located within an atom. How can that motion be explained?

1.2. Explain why water flies off a wet dog that is shaking himself.

1.3. Explain why you slide toward the outside of the curve when sitting in a car going around a curve at high speed.

1.4. A car is moving at 60 miles/hour. Express this speed in kilometers per second. (0.03 km/s)

1.5. A person weighs 155 pounds. Express this in kilograms. (70.3 kg)

1.6. An air conditioner is rated at 5000 BTU per hour. Express this in Joules per second. (1465 J/s)

1.7. The motor of a car may produce 75 horsepower. Express this in kilocalories per second. (13.4 kcal/s)

1.8. The mean distance from the earth to the sun is 93×10^6 miles. Express this in kilometers. (150×10^6 km)

1.9. An electric heater is advertised to produce heat at the rate of 1500 watts, and a gas heater is advertised to produce heat at the rate of 8000 BTU per hour. Which of these actually produces heat faster? (P_{gas} = 2344 W, $P_{electric}$ = 5119 BTU/h, so the gas heater produces heat faster than the electric heater does.)

1.10. A car is moving at 45 miles/hour and then accelerates to 60 miles/hour in 3 seconds. Express the acceleration of the car in g's. (0.23 g's)

1.11. A 175-lb pilot flying a jet plane experiences an acceleration of 4 g's. How many pounds of force are acting on her? (700 lb)

1.12. A car is advertised to be capable of accelerating from 0 to 60 miles per hour in 9 seconds. Assume that the acceleration was constant, and determine
a. the acceleration in g's. (0.3 g's)
b. the average velocity (30 miles/hour)
c. the distance covered (396 ft)

1.13. It is known that a person's upper body (trunk, head, and arms) weighs 85 lb. Consider such a person who is standing erect. How much force do the person's hips, which are supporting the upper body, exert? (85 lb, up)

1.14. The deltoid muscle produces a force of 75 lb pulling the upper arm (humerus) into the shoulder. How much force does the shoulder exert on the humerus? (75 lb, pushing out)

1.15. A 75-kg person is standing on the floor. How much force does the floor exert on the person? (735 N, up)

Vectors

Introduction to Vectors

To continue the development of a quantitative description of motion, we must introduce an important mathematical quantity: the vector. We saw in the previous material that the direction associated with velocity is important and plays a role in explaining what is going on. This is so common that a special category of quantities has been identified to represent this characteristic. A **vector** is a quantity that requires both a magnitude (numerical size and units) and a direction for a complete description. A vector will be noted in the text as a symbol with an arrow over it. Among the vectors that we shall use in this course are velocity, acceleration, and force. So, for example, velocity will be denoted as $\vec{V}$ and force as $\vec{F}$. If the velocity of an object changes, we will denote the various velocities by the use of subscripts, such as $\vec{V}_1$ and $\vec{V}_2$; different forces might also be distinguished by the use of subscripts, such as. $\vec{F}_1$ and $\vec{F}_2$. The vector quantity that we will encounter most often in this course is force, and so it will be used as the basis of most of the examples. However, it is important to note that the vector techniques that are described below apply to all vectors, not only forces. (*Note*: There are other quantities that do not involve direction; they are called scalars. A **scalar** is a quantity that requires only a magnitude (numerical size and units) for a complete description. Examples of scalars that we will use in this course are time, mass, energy, temperature, and heat.)

As was mentioned above, the vector quantity that is most encountered in this course is force. This should not be surprising. We affect objects in the world around us by exerting forces on them. If you want to throw a ball, you push forward on the ball. If you want to lift something, you exert an upward force on it. Suppose that you wanted to lift a heavy rock out of a hole. If you bent down and tried to pull it out, you might hurt your back. There is another way. You could tie a rope around the rock, pass the rope up and over a wheel (perhaps located in a tree), and then pull down on the rope. Although you are pulling down on the rope, the rope pulls up on the rock. The wheel serves to change the direction of the force. Such a wheel is called a pulley. We could have passed the rope over a branch of the tree rather than a pulley. That would have worked, but it would have been more difficult because of the increase in friction where the rope rubs on the branch. Similar situations occur within your body. Consider one of your knees.

Suppose that you want to kick forward with your lower leg, for example kicking a soccer ball. If you want your leg to accelerate forward, there must be a force directed forward that acts on the leg. This force could be either a push from behind or a pull from in front. As we have already mentioned, soft tissue (muscle, tendon, etc.) can only pull. It cannot push. So there must be some tissue that can pull the lower leg forward. The muscle that causes the lower leg to kick forward (to extend) is the quadriceps. This muscle is located on top of the thigh, above and behind the lower leg. So how can it pull the lower leg forward? The tendon that connects the quadriceps to the tibia (the lower leg) passes over the patella (the kneecap). The patella acts like a pulley, changing the direction of the force. So although the quadriceps pulls the tendon up and to the left, the tendon pulls the tibia up and to the right.

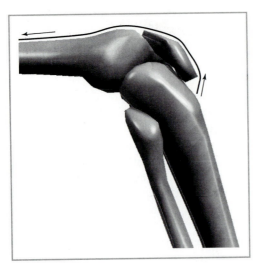

An important characteristic of pulleys is that the tension is equal everywhere in whatever (rope or tendon) is passing around the pulley. Thus, in the figure above, the tension is equal in the part of the tendon above the patella and in the part of the tendon below the patella. The directions are different, but the magnitudes are equal.

Pulleys play an important role in the traction setup for a person with a broken leg. The rope passing from the leg to a hanging weight may pass over several pulleys. The tension in the rope is the same everywhere.

Addition of Vectors

Consider a student who is sitting at a desk, listening to a lecture. It is highly likely that he will be leaning forward, one of his hands supporting his chin. This is automatic, but why does it happen? From Galileo's work on the relation between force and motion, we know that if a force acts on an object, then that object will accelerate. That is, it will speed up, slow down, or change direction. It is reasonable to assume that the student's head is not accelerating. Thus we know that the total force acting on his head must equal zero. What forces are acting? There would be

the force of gravity pulling his head down,

the force of the topmost cervical vertebra (C1) pushing up and forward,

the splenius muscles, which originate on the cervical and thoracic vertebrae and insert onto the skull, thus pulling the head down and backward and

the force of the hand pushing upon the chin:

$$\vec{F}_{resultant} = \vec{F}_{hand} + \vec{F}_{gravity} + \vec{F}_{vertebrae} + \vec{F}_{muscles}$$

As we stated above, the total of these four forces (also called the net force or resultant force) must be zero because the acceleration of the head is zero. To carry out the analysis, we must know how to add forces together. When adding forces, we must deal with the fact that

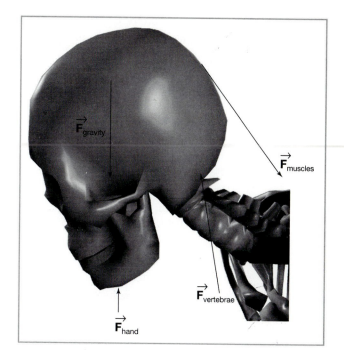

force, such as velocity and acceleration, is a vector, and so we must use **vector addition**. There are two ways of adding vectors: a graphical technique and an analytic technique.

The **graphical technique** involves using a ruler and a protractor to make a scale drawing of the vectors. The magnitude of the resultant vector is then measured by using the ruler and the direction is measured by using the protractor.

The **analytical technique** involves using algebra and trigonometry to represent the vectors mathematically. They can then be added together, and the magnitude and direction of the resultant can be calculated.

The graphical approach is more visual, but errors may be introduced by inaccuracies in making the drawing or making measurements. The analytical approach is not subject to this type of error and is therefore more accurate. It is also more powerful because, since it is based on calculations, it may employ programmable calculators and computers.

Graphical Technique

To use the graphical technique, first assume that we know the magnitudes and directions of the vectors that we wish to add and/or subtract.

The procedure for graphical combination of vectors is as follows:

1. Start with a vector equation in which the unknown vector appears alone on the left side.

2. Choose a scale such that the arrows representing the vectors will fit on the paper. It is also very important that the resulting drawing be large so that accurate measurements can be made from it; for example, the scale could be 20 units = 1 inch, 1 cm = 5 units, 1 cm = 5 lb, 1 cm = 5 N, or 1 cm = 5 m/s.

3. Choose a point to be the origin, and draw a set of mutually perpendicular axes. By convention, the X axis is drawn horizontally and the Y axis is drawn vertically. We will always orient the +Y axis to be 90° counterclockwise from the +X axis.

4. Choose a direction that will be 0°. A standard convention that is used in dealing with directions is to assign 0° to the +X axis, usually drawn pointing to the right. Angles are then measured counterclockwise from the +X axis.

5. Pick one of the vectors, and draw it pointing outward from the origin of the coordinate system. (*Note*: If the vector is negative, it is drawn in a direction that is 180° from its specified direction.)

6. Draw a new origin at the head of this vector.

7. Pick one of the other vectors, and draw it pointing outward from this new origin.

8. Draw a new origin at the head of this vector.

9. Pick one of the other vectors, and draw it pointing outward from this new origin.

10. Continue this process until all of the vectors have been used.

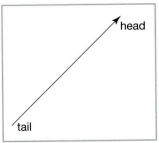

11. The unknown vector can now be drawn. It will point from the *original* origin to the head of the last vector drawn.

12. The magnitude of the unknown vector can be determined by measuring its length and then using the scale that was specified in Step 1.

13. The direction of the unknown vector can be determined by using a protractor.

Example 1.14

A child is pulling a 10-lb box by exerting a force at an angle of 30° above the horizontal. Let us assume that there is a frictional force of 2 lb acting between the floor and the bottom of the box. The problem is to determine how much force the child is applying such that the box moves at constant speed.

Solution

1. What forces are acting on the box?
   a. The child is pulling up and to the left.
   b. Gravity is pulling the box straight down.
   c. The ground is pushing straight up.
   d. There is a force of friction acting to the right (opposite to the direction of the box's motion).

   The four forces are sketched as arrows (page 31).

2. Since the box is moving at constant speed, the acceleration is zero.

3. Therefore the total force (the sum of the forces) must equal zero.

4. We can write the following vector equations:

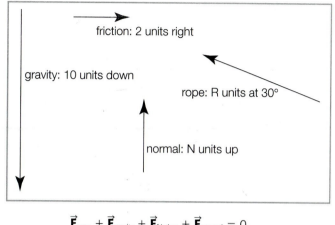

$$\vec{\mathbf{F}}_{rope} + \vec{\mathbf{F}}_{gravity} + \vec{\mathbf{F}}_{friction} + \vec{\mathbf{F}}_{normal} = 0$$
$$\vec{\mathbf{F}}_{rope} = -\vec{\mathbf{F}}_{gravity} - \vec{\mathbf{F}}_{friction} - \vec{\mathbf{F}}_{normal}$$

We now have a vector equation in the proper form for the application of the graphical technique.

5. The graphical technique may now be used to determine the amount of force exerted by the child on the box.

a. An origin is chosen, and a set of axes is drawn.

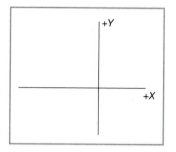

b. From the origin, one of the vectors $(-\vec{\mathbf{F}}_{gravity})$ is drawn. Notice that it is pointing up rather than down. The force of gravity always points down, but the negative sign means that it is drawn pointing up.

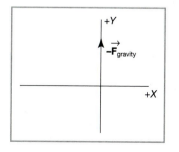

c. From the end of this vector, another one $(-\vec{\mathbf{F}}_{\text{friction}})$ is drawn.

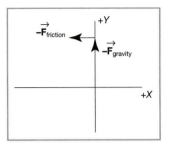

d. From the head of this vector, another one $(-\vec{\mathbf{F}}_{\text{normal}})$ is drawn. The length of the vector is not known, but since its direction is (a normal force always points perpendicular to the surface and toward the object; thus it would point up but the negative sign in front of it changes that direction), we can draw a line pointing downward.

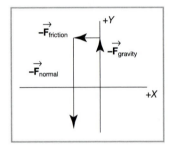

e. We have now drawn the three vectors that appear on the right-hand side of the equation. The resultant vector, $\vec{\mathbf{F}}_{\text{rope}}$, is now drawn. Once again, we do not know its length, but we do know its direction. A line may be drawn from the original origin directed at 30° above the horizontal. Since we know that this vector must terminate at the head of the last of the three vectors previously drawn, the arrowhead is located, and the magnitude of the vector may be measured.

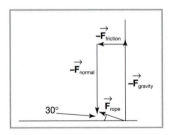

f. The magnitudes of the normal force and of the force of the rope can now be determined.

Analytical Technique

The analytical technique is based on calculations rather than a scale drawing. Before we get into the details of the analytical technique, it is necessary to introduce a new concept: **com-**

ponents of a vector. We have already seen that, to completely describe a vector, it is necessary to give its direction as well as its magnitude. The magnitude will be a number with units, such as 25 miles per hour or 456 Newtons. Specifying the direction is a bit more complicated. We could use angles, such as 20° below the $+X$ axis, 35° clockwise from the $-Y$ axis, or "straight down." This is not the only way to specify a vector.

Suppose that you were in a city and were asked for directions. Typically, you would not give a distance and direction for the person to walk because buildings will probably be in the way. Your directions would involve walking a certain number of blocks in a particular direction, then making a 90° turn and walking a certain number of blocks in that direction. This is the approach that we will use.

EXAMPLE 1.15

We could specify both the magnitude and direction of a position vector ($\vec{P}$) by saying that it is 250 feet long and points at 55° to the east of due north. However, we could also say that the position is 205 feet east and 143 feet north of the starting point.

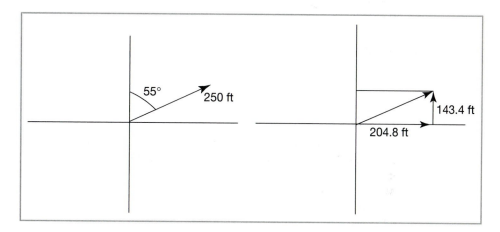

The latter approach involves the components of the vector. Notice that one component is along the X axis and the other is along the Y axis. Referring to the diagram on the right, we would say that the X component of the vector is 204.8 ft and that the Y component of the vector is 143.4 ft. By convention, the X component of a vector is always parallel to the X axis and the Y component is always parallel to the Y axis.

$$P_x = 204.8 \text{ ft}$$
$$P_y = 143.4 \text{ ft}$$

Although this example refers to a position vector, components may be used for any kind of vector, for example, force or acceleration.

There are two techniques, either of which may be used to determine the components of a vector:

Technique 1: Draw a vector as an arrow pointing outward from the origin. The components of a vector are given by the coordinates of the endpoint of the arrow. The components are calculated by using the sides of a right triangle and the basic trigonometric functions: sine, cosine, and tangent.

For example, consider a vector   that has a magnitude of 8 N and points   counterclockwise (CCW) from the +Y axis. Notice that $\vec{F}_1$ points up and to the left.

Its X and Y components can be found as follows:

$$F_{1x} = -8 \cos 60°$$
$$= -4$$

(*Note*: In using a calculator to determine the value of a trig function, e.g., cos 60°, make sure that the calculator is in the degree mode and not the radian mode.)

Note that **this must be negative** because the X component of the vector $\vec{F}_1$ points to the left and the standard sign convention is that the + X direction is to the right.

$$F_{1y} = 8 \sin 60°$$
$$= 6.93$$

This will be positive because the Y component of $\vec{F}_1$ is pointing up, and up is the standard direction for +Y.

Remember that it will be necessary (see the alternative technique described below) to label the components as + or − by using a standard convention: An X component that points to the right is +, and one that points left is −; a Y component that points up is +, and one that points down is −.

Technique 2: Alternatively, one may avoid the step of dealing with the individual triangles as follows: **If all of the angles are measured counterclockwise from zero degrees on the positive X axis, then an X component will always involve the cosine of the angle, and a Y component will always involve the sine of the angle.** An advantage to this approach is that minus signs will appear automatically and need not be considered separately as in the method described above.

$$\text{Vector}_x = (\text{vector magnitude}) \cos (\theta) \qquad V_x = V \cos (\theta)$$
$$\text{Vector}_y = (\text{vector magnitude}) \sin (\theta) \qquad V_y = V \sin (\theta)$$

EXAMPLE 1.16

Consider again an 8-lb force directed at 30° CCW from the +Y axis. We want to determine the X and Y components of this force. Notice that this force points up and to the left.

Solution

It is not necessary to draw any triangles. We use the following formulas:

$$F_x = F \cos (\theta) \qquad \text{and} \qquad F_y = F \sin (\theta)$$

Following the convention described above, the angle θ must be 120°:

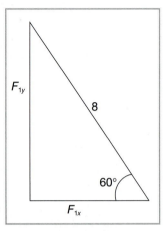

$$F_x = 8 \cos 120 \qquad F_y = 8 \sin 120$$
$$F_x = -4 \text{ lb} \qquad F_y = 6.93 \text{ lb}$$

Notice that the *X* component comes out negative automatically.

EXAMPLE 1.17

The figure on the left shows a man holding a ball out to the side. The figure on the right shows his left arm. The heavy black line in the figure on the right represents the tendon of the deltoid muscle. When this muscle, located at the shoulder, contracts, it pulls on the tendon, which, in turn, pulls on the upper arm. Thus the tendon in the picture above pulls the arm up and to the left.

A sketch of this force is shown in the graph. A reasonable value for the angle *A* is 12°.

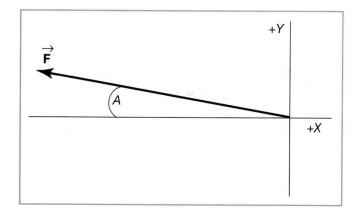

The *X* component will be given by:

$$F_x = F \cos 168°$$

It will **cause compression of the shoulder** (because it points directly at the shoulder joint). The *y* component will be given by:

$$F_y = F \sin 168°$$

It will **cause rotation of the arm about the shoulder joint** (because it is perpendicular to the upper arm).

EXAMPLE 1.18

The figure shows a woman lying on a floor while doing an exercise (intended to build up her hip abductor muscle) and a magnified view of her left hip. The hip abductor muscle applies tension between the greater trochanter of the femur and the ilium. It pulls the greater trochanter down and to the left. The angle between the neck of the femur and the femur itself is 125°, and the angle between the hip abductor muscle and the axis of the femur is 8°. **(Note: In working through this example and other anatomy-based examples and problems, see Appendix 9 for detailed information about anatomy.)**

Consider only the force (assumed to be 104 lb) represented by the pull of the hip abductor muscle and determine the following:

1. How many pounds are compressing the femur-acetabulum joint?
2. How many pounds are causing rotation of the femur?

3. Consider a person who is suffering from coxa valga, in which the angle between the femur and the neck of the femur is larger than normal, perhaps 160° rather than 125°. In this case, what would have to be the tension in the hip abductor muscle to produce the same amount of force tending to rotate the femur as in question 2?

Solution

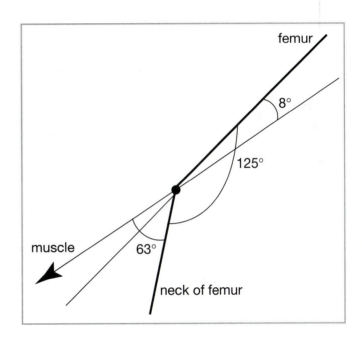

The important idea to be used in this example is that if a force acts on a bone, the component that lies along the bone will cause compression or (rarely) dislocation of the joint, while the component that is perpendicular to the bone will cause rotation about the joint. Thus to answer question 1, we must determine the magnitude of the component of the 104-lb force that lies along the neck of the femur. In question 2, we must calculate the magnitude of the perpendicular component.

We want to determine the component that is along the neck of the femur (to calculate the compressive effect of the muscle). Recall that components are, by their basic definition, measured along the coordinate axes. Since we want to determine components that are either parallel to or perpendicular to the neck of the femur, these two directions must be used to draw the X and Y axes. Let us label the neck of the femur to be the +X axis. Since the +Y axis must be 90° CCW from the +X axis, we label it as shown in the next figure.

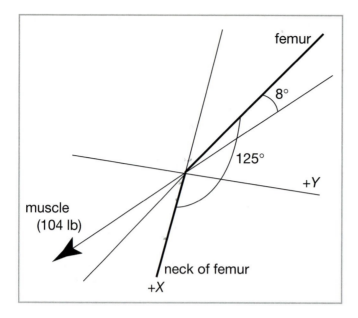

We may now use the formulas established earlier:

$$F_{muscle,x} = F_{muscle} \cos \theta$$
$$F_{muscle,y} = F_{muscle} \sin \theta$$

To use these formulas, we need the angle that is labeled as A in the diagram.

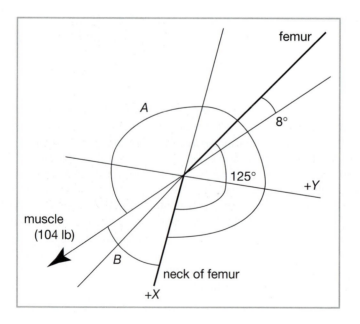

The angle between the $+X$ axis and the extended muscle line is equal to $125° - 8° = 117°$. This means that the angle between the $+X$ axis and the actual muscle vector (angle B) is $180° - 117° = 63°$. Therefore the angle $A = 360° - 63° = 297°$.

Substituting, we determine that the compressive effect of the 104-lb force (its component along the neck of the femur) is $104 \cos 297° = 47.2$ lb (answer to question 1) and the rotation effect (the component perpendicular to the neck of the femur) is $104 \sin 297° = -92.7$ lb (answer to question 2). (The $-$ sign indicates the direction.)

If the angle between the femur and the neck of the femur were 160°, the angle between the hip abductor muscle and the neck of the femur would become 28°. Proceeding as above, we determine that

$$F_{rotation} = F_{muscle,y} = F_{muscle} \sin \theta$$
$$92.7 = F_{muscle} \sin 332°$$
$$F_{muscle} = 197.5 \text{ lb (answer to question 3)}$$

This means that a person who suffers from coxa valga, a larger than normal angle between the neck of the femur and the axis of the femur, must have a far greater tension in the hip abductor muscles to achieve the same amount of rotation of the legs about the hip. This, of course, will result in more compression of the femur-acetabulum joint.

EXAMPLE 1.19

An understanding of the components of a vector can be very important. Consider the forces that act on one of the lumbar vertebrae. If the vertebrae were stacked vertically like a stack of checkers,

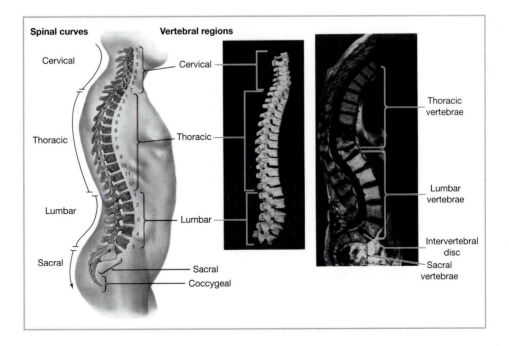

the force on one of them from the top would be straight down, and the force on it from the bottom would be straight up. Thus the only effect on the vertebra would be compression. Since bone is extremely strong in withstanding compression, there would be little or no resulting damage.

Unfortunately, this is not the real situation. The spinal column in the lumbar region is not straight but is curved. The bearing surfaces of the vertebrae are not parallel horizontal planes but are inclined. Thus the force from above (the weight of the upper body that is directed straight down) is not perpendicular to the bearing surfaces and therefore will not have a solely compressive effect.

We may analyze the effects of this force by considering its components. The force may be represented by two components: one perpendicular to the surface and one parallel to the surface. The perpendicular (also called the normal) component will cause compression that is not dangerous. However, the parallel (also called the shear) component will cause the bones to tend to slip past each other. This potential motion is constrained not by bone but by the soft tissue—that is, the cartilage—that surrounds the spinal column. This material may not be capable of withstanding the shearing effect. If the vertebrae slip or if the cushioning material (the discs) between the vertebrae is distorted, pressure on the skeletal nerves that pass out from the spinal column may result. This can produce lower back pain and possibly partial paralysis of one of the legs.

EXAMPLE 1.20

The sketch represents the lowest lumbar vertebra (L5) and the sacrum (S1) of a 150-lb person. F_1 is the total force exerted by L5 on the sacrum; F_c is the component of F_1 that is compressing the sacrum (the normal or perpendicular component); and F_s is the component of F_1 that will cause slippage at the joint (the parallel component).

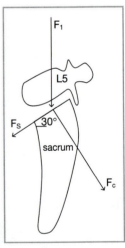

1. How much force is represented by the symbol F_1?
2. How much force is represented by the symbol F_s?
3. How much force is represented by the symbol F_c?

Solution

The weight of the person is given as 150 lb. However, not all of this weight pushes down on L5 and S1. Only the weight of that part of the body above L5 and S1 pushes down on the joint. Using the chart relating to the weights of body segments in Appendix 9, we see that 66.1% of the body weight is above L5 and S1. Therefore F_1 will be 0.661 × 150 lb ... F_1 = 99 lb (answer to question 1). (**Note: We will later learn that the weight of the parts of the body that are above L5 and S1 play only a small part in the amount of force compressing that joint. Most of the force of compression is due to the tension in the muscles of the back. See Example 2.12 relating to the lower back in the section on torque in Chapter 2.)**

We now have the situation depicted in the graph on page 41 (left).

Now we need a set of axes. Usually, these are drawn horizontal and vertical. However, in this problem, we do not want horizontal and vertical components; we want components that are along the two lines indicated in the graph on page 41 (left). So we may rotate the axes until they

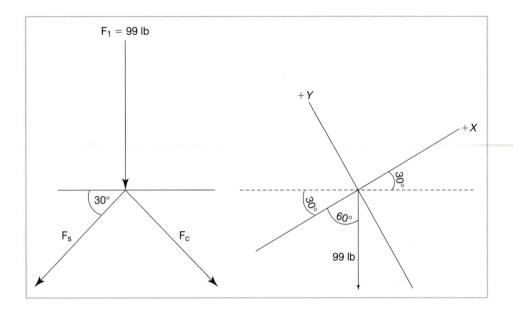

align with the directions of the desired components. The known vector is then drawn on this set of axes, as shown in the graph above (right).

Using the second technique, described above, we would have

$$F_x = 99 \cos 240° = -49.5 \text{ lb (answer to question 2)}$$
$$F_y = 99 \sin 240° = -85.8 \text{ lb (answer to question 3)}$$

Thus there will be a force of 85.8 lb compressing the joint and a force of 49.5 lb that will have the effect of making the bones slip past each other. The actual bony material of the sacrum is used to withstand the 85.8 lb force, and the soft tissue (ligaments and cartilage) that stabilize the L5-S1 joint serve to withstand the 49.5 lb force. Since bone is very strong against compression, the sacrum can easily cope with the 85.8-lb force. However, if the angle between the bearing surface of the sacrum and the horizontal were 60° rather than 30°, F_c would equal 49.5 lb and F_s would equal 85.8 lb. Now the soft tissue around the joint must withstand the larger force. This might lead to eventual failure of the soft tissue.

EXAMPLE 1.21

Consider the forces that act on the upper end of the tibia when it is partially flexed (see page 28). The quadriceps tendon is pulling up and to the right. There is a component (directed toward the tibia-femur joint) that will serve to compress the joint while the perpendicular component will produce a shear. It will cause the top of the tibia to slide past the end of the femur. This shear would have the effect of dislocating the knee and must be countered. The anterior cruciate ligament (ACL) is located within the joint, connecting the front of the upper surface of the tibia to the rear of the lower surface of the femur. Thus, the ACL has the effect of pulling the top of the tibia up and backward. It is the tension in this tendon that stabilizes the knee against the dislocating effect of the shear component of the tension in the quadriceps tendon.

EXAMPLE 1.22

In our discussion of friction (see page 25), we saw that a person will automatically take smaller steps when walking on a slippery surface. We can now better analyze that action. Consider a woman whose legs are 75 cm long. If she is walking on a slippery sidewalk ($\mu_s = 0.2$), what is the longest stride that she can take without slipping?

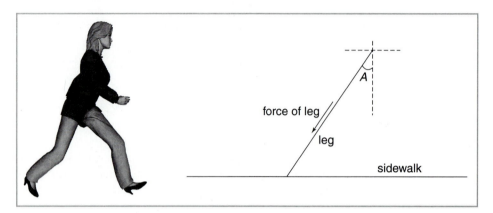

The force of the rear leg F_l has two components: F_{LV} down and F_{LH} backward. Considering the lower triangle, we have

$$F_{LV} = F_L \cos A$$
$$F_{LV} = F_L \sin A$$

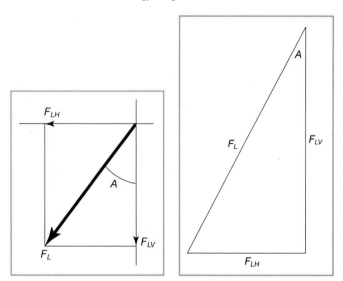

We shall see later that if only one foot is in contact with the ground at this instant, the vertical component of the force exerted by the leg will be equal to the woman's weight:

$$F_L \cos A = mg$$

This allows us to calculate the magnitude of the force exerted by the leg:

$$F_L = \frac{mg}{\cos A}$$

We may now substitute this quantity into the expression for F_{lh}:

$$F_{LH} = mg \tan A$$

In order that the woman's foot does not slip, it is necessary that F_{LH} be less than the force of static friction:

$$F_{LH} < \mu_s F_n$$
$$mg \tan A < \mu_s F_n$$

The magnitude of the normal force, F_n, is equal to her weight, mg, and so

$$\tan A < \mu_s$$

Therefore, to keep from slipping, she takes small steps, keeping the angle A very small.

The question originally posed was to determine the length of the longest stride that a person could safely take on a slippery sidewalk. The length of the stride is

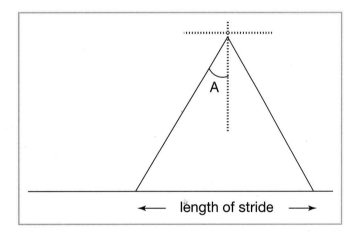

$$L_{stride} = 2L_{leg} \sin A$$
$$\tan A < \mu_s$$
$$A < \tan^{-1}(\mu_s)$$
$$\frac{L_{stride}}{2L_{leg}} < \sin(\tan^{-1}\mu_s)$$
$$L_{stride} < 2L_{leg} \sin(\tan^{-1}\mu_s)$$
$$L_{stride} < 2(75) \sin(\tan^{-1}(0.2))$$

Thus the maximum length of the woman's stride will be 29 cm.

We see from these examples that components of a force are not abstract mathematical quantities. They can be important in understanding a basic part of the functioning of the body.

Now that we have a good idea of what is meant by the components of a vector, we can discuss a procedure that uses components to combine (add or subtract) different vectors. The following procedure represents a worst-case scenario; you might not need all of these steps to solve a specific problem. For example, we will assume that the vectors are forces (the usual case in this course). If they are not forces, Steps 1, 2, 3, and 6 are omitted.

EXAMPLE 1.23

With reference to the diagram of the knee, the pull or tension in the quadriceps (or transpatellar) tendon is 300 lb. The angle between the tendon and the horizontal above the knee is 37°, and the angle between the tendon and the horizontal below the knee is 80°. (*NOTE*: The force of gravity (the weight) acting on the patella may be ignored in this problem.) Determine the magnitude and direction of the force $\vec{F}_c$ that the femur exerts on the patella.

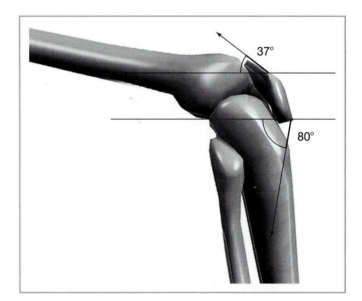

Solution

In this solution, there are two sets of steps: one in italics and one in the normal font. The italicized steps are general ones that may be used in many different problems. The normal font steps are specific to this problem.

1. *Identify the object of interest.*
1. Since there are several different bones in this problem, we must select one of them to serve as the object of interest. This will be the object on which the various forces are acting: the patella.

2. *List the forces that act **on that object.***
2. We then list the external forces: those that act on the object. The forces acting on the *patella* are as follows:

Force of the upper tendon: $\vec{\mathbf{F}}_u$ = 300 lb to the left at 37° above the horizontal.
Force of the lower tendon: $\vec{\mathbf{F}}_l$ = 300 lb to the left at 80° below the horizontal.
Force of the femur: $\vec{\mathbf{F}}_c$ is unknown.

3. *Make a free body diagram (FBD).*
3. The FBD is now constructed. A dot is drawn to represent the object of interest. Arrows emanating from the dot are drawn to represent the external forces. In this example, $\vec{\mathbf{F}}_c$ represents a difficulty because we do not know its direction. I suggest that when faced with drawing a vector whose direction is unknown, you always draw it pointing into the first quadrant (usually up and to the right). This will result in its having positive X and Y components in the equations. If the vector actually has one or the other as negative components, this will work out mathematically.

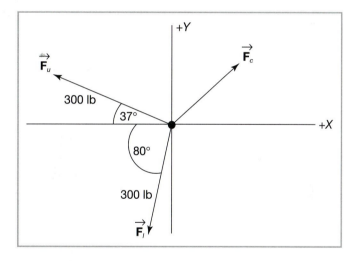

4. *Write the basic equation in vector form.*
4. Since the patella is not accelerating, the sum of the forces acting on it must equal zero:

$$\sum \vec{\mathbf{F}}_{\text{external}} = 0$$

$$\vec{\mathbf{F}}_u + \vec{\mathbf{F}}_l + \vec{\mathbf{F}}_c = 0$$

5. *Write the equivalent scalar equations.*
5. The equivalent scalar equations are:

$$\sum F_{\text{external}, X} = 0 \qquad F_{uX} + F_{lX} + F_{cX} = 0$$

$$\sum F_{\text{external}, Y} = 0 \qquad F_{uY} + F_{lY} + F_{cY} = 0$$

6. *Substitute from the FBD.*
6. We have

$$X \text{ equation:} \quad 300 \cos 143 + 300 \cos 260 + F_{cX} = 0$$
$$Y \text{ equation:} \quad 300 \sin 143 + 300 \sin 260 + F_{cY} = 0$$

7. *Solve for the unknowns.*
7. We have

$$F_{cX} = 291.7 \text{ lb}$$
$$F_{cY} = 114.9 \text{ lb}$$

Notice that both F_{cX} and F_{cY} came out to be positive numbers. This means that our assumption that $\vec{F}_c$ points up and to the right was correct.

8. *Represent the calculated components on a set of coordinate axes.*

   a. *Draw the rectangle bounded by the components.*
   b. *Draw the vector pointing from the origin to the opposite corner of the rectangle.*

8. For this example we have this sketch:

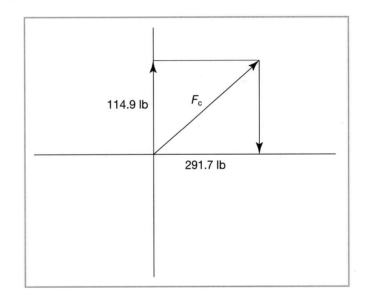

9. *Draw one of the resulting right triangles, labeling each side and any known angles.*
9. For this example, we have this sketch:
10. *Use trigonometry to solve for the magnitude and direction of that vector.*

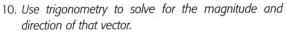

10. Solve

$$F_c = \sqrt{(291.7)^2 + (114.9)^2}, \quad F_c = 313.5 \text{ lb}$$

$$A = \tan^{-1}\left(\frac{114.9}{291.7}\right), \quad A = 21.5° \text{ CCW from the } +X \text{ axis}$$

EXAMPLE 1.24

Refer to Example 1.14, regarding a child pulling a box.

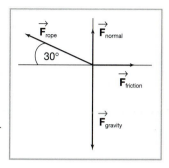

Step 1. The object of interest is the box.

Step 2. The forces acting on the box are as follows:

    a. the child pulling up and to the left,
    b. gravity pulling the box straight down, and
    c. the ground pushing straight up.
    d. there is a force of friction acting to the right (opposite to the direction of the box's motion).

Step 3. The FBD on the right:

Step 4. The basic equation: Write the basic equation in vector form. In this example, the box is not accelerating, and therefore the sum of the external forces must be zero.

$$\sum \vec{F}_{external} = 0$$

$$\vec{F}_{rope} + \vec{F}_{gravity} + \vec{F}_{friction} + \vec{F}_{normal} = 0$$

$$\vec{F}_{rope} = -\vec{F}_{gravity} - \vec{F}_{friction} - \vec{F}_{normal}$$

Step 5. Scalar (or component) equations: Each vector is replaced by it scalar or component equations, such as:

$$F_{rope,X} = -F_{gravity,X} - F_{friction,X} - F_{normal,X}$$
$$F_{rope,Y} = -F_{gravity,Y} - F_{friction,Y} - F_{normal,Y}$$

Step 6. Determination of the components: Each of the terms in these equations is called a component of the corresponding vector. So, for example, $F_{rope,X}$ is the X component of the vector $\vec{F}_{rope}$.

$$F_{friction,X} = 2 \cos 0° = 2 \qquad F_{friction,Y} = 2 \sin 0° = 0$$
$$F_{gravity,X} = 10 \cos 270° = 0 \qquad F_{gravity,Y} = 10 \sin 270° = -10$$
$$F_{normal,X} = F_n \cos 90° = 0 \qquad F_{normal,Y} = F_n \sin 90° = F_n$$
$$F_{rope,X} = F_r \cos 150° \qquad F_{rope,Y} = F_r \sin 150°$$

Step 7. Substitution into the scalar equations:

$$F_{rope,X} = -F_{gravity,X} - F_{friction,X} - F_{normal,X}$$
$$F_R \cos 150° = 0 - (2) - 0$$
$$F_R = 2.31$$
$$F_{rope,Y} = -F_{gravity,Y} - F_{friction,Y} - F_{normal,Y}$$
$$F_{rope} \sin 150° = -(-10) - 0 - F_{normal}$$
$$(2.31)(\sin 150°) = 10 - F_{normal}$$
$$F_{normal} = 8.85$$

Notice that F_{normal} comes out to be a positive number. That means that the arrow representing F_{normal} was correctly drawn on the FBD. If the calculated component had come out to be a negative number, it would mean that we had drawn that component pointing in the wrong direction on the FBD. It would really point in the direction opposite to the way it was drawn.

We see that the child is pulling with a force of 2.31 lb and that the ground is pushing up on the box with a force of 8.85 lb.

Notice the emphasis on forces acting on the object. This cannot be overemphasized. It will always be the external forces (those acting on the object) that are important in the analysis. There may be internal forces (those acting within the object) and forces that the object exerts on something else. Neither of these two types of forces will play a role in the analysis.

Consider your body as you are seated at the desk. You are not accelerating, so it is fair to say that the forces acting on your body must add up to zero. There are certainly forces acting within your body. Your heart is pushing blood up to your brain, many muscles are in tension pulling on tendons, which are, in turn, pulling on bones, and so on. However, because these forces are internal, they have no effect on the acceleration of your body. The force of gravity from the Earth is pulling your body down, and so your body is pushing down onto the chair. Since this force is exerted by your body rather than on your body, it will not play a role in the body's acceleration. However, there is a related force that does play a role: the force exerted by the chair on your body (an external force). These two forces, the one exerted by your body on the chair and the one exerted by the chair on your body, are related to each other by Newton's third law.

Since your body is pushing down on the chair, the chair is pushing up on your body. Thus the external forces acting on your body are the force of gravity pulling you down and the force of the chair pushing you up.

We unconsciously make use of Newton's third law when we stand on a bathroom scale. Consider a woman standing on a bathroom scale. The force of gravity (due to the Earth) is pulling her down, and as a result, she is pushing down on the scale. The scale is pushing up on her. Thus the woman and the scale are interacting. They are exerting forces on each other. Because of the Third Law, we can say that these two forces are equal in magnitude and opposite in direction. The "opposite in direction" part should be obvious. The mechanism in the scale is a spring with a pointer attached. As she pushed down on the scale, the spring compresses and the pointer moves. The pointer is actually measuring the force exerted by the scale. Because of the Third Law, we associate this with her weight. If she were to push down on the scale with a force greater than her weight, the reading on the scale would also be greater than her weight. For example, she might stand on the scale, crouch down, and then jump up. While she was jumping up off the scale, the reading on scale would be much greater than her weight. As another example, she might take the scale into an elevator and stand on it as the elevator went up. If she were to look at the scale's reading as the elevator came to a stop, she would find that she seemed to have lost weight. Alternatively, if she were to look at the reading as the elevator came to a stop after moving down to a lower floor, she would find that she seemed to have put on weight. We will discuss these situations in the next section.

Now that we have discussed vectors and components, we can return to the mathematical description of motion that we started in the previous section. There, we defined acceleration both conceptually and through a vector equation:

$$\vec{a} = \frac{\Delta \vec{v}}{\Delta t}$$

This equation can be expressed in terms of components:

$$a_X = \frac{\Delta v_X}{\Delta t} \quad \text{and} \quad a_Y = \frac{\Delta v_Y}{\Delta t}$$

We will now use the position vector ($\vec{P}$), which we have already introduced, to define the displacement vector, $\vec{D}$. This vector describes the motion of an object in that it points from an earlier position of the object to a later position. The magnitude of $\vec{D}$ is the straight-line distance between the initial and final positions. Mathematically,

$$\vec{D} = \vec{P}_2 - \vec{P}_1$$

We can express its X and Y components as

$$
\begin{array}{lll}
D_X = P_{2X} - P_{1X} & \text{or} & \Delta X = X_2 - X_1 \\
D_Y = P_{2Y} - P_{1Y} & \text{or} & \Delta Y = Y_2 - Y_1
\end{array}
$$

Once we have the displacement vector defined, we can define the average velocity:

$$\vec{\bar{v}} = \frac{\vec{D}}{\Delta t}$$

This vector will have components defined by

$$\bar{v}_X = \frac{\Delta X}{\Delta t} \quad \text{and} \quad \bar{v}_Y = \frac{\Delta Y}{\Delta t}$$

The last vector equation that we shall need for a mathematical description of motion is

$$\vec{\bar{v}}_X = \frac{1}{2}(\vec{v}_1 + \vec{v}_2)$$

(*Note*: **This equation is valid only for situations in which the acceleration is constant. However, that will cover all of the examples and problems in this text.**)

In terms of components, we will have

$$\bar{v}_X = \frac{1}{2}(v_{1X} + v_{2X}) \quad \text{and} \quad \bar{v}_Y = \frac{1}{2}(v_{1Y} + v_{2Y})$$

It is important to note that displacement is not the same as distance. To illustrate this, we will continue Example 1.4 (page 9) relating to a ball that is thrown straight up into the air with an initial velocity of 40 ft/s. We will use the Y component equations for average velocity and displacement above.

t_2 (S)	v_{2Y} (ft/s)	$\bar{v}_Y$ (ft/s)	ΔY (ft)
0	40	not defined	0
0.25	32	36	9
0.5	24	32	16
1.0	8	24	24
1.25	0	20	25
1.5	−8	16	24
1.75	−16	12	21
2.0	−24	8	16
2.25	−32	4	9
2.5	−40	0	0

We have already shown (page 10) that the ball reaches its maximum height at the end of 1.25 s. From the table, then, we see that the maximum height of the ball is 25 ft. If it travels 25 ft going up, then it must travel 25 ft coming back down. The distance the ball travels is clearly 50 ft. However, from the table, we see that the displacement of the ball at the end of the trip (2.5 s) is zero. **So displacement and distance are not the same.** Just as velocity was defined in terms of displacement and time, speed is defined in terms of distance and time. Since velocity and distance are not the same, it follows that **velocity and speed are not the same.** If we were to try to define the speed of the ball as distance divided by the time interval:

$$\text{speed} = \frac{\text{distance}}{\text{time interval}},$$

we would get (for the whole trip)

$$\text{speed} = \frac{50 \text{ ft}}{2.5 \text{ s}} = 20 \frac{\text{ft}}{\text{s}}.$$

Although this number is correct in terms of arithmetic, it has no meaning. It does not represent how fast the ball is moving as it hits the ground, how fast it is moving at the top of its path, or how fast it is moving at any other special time. The basic issue that makes this calculation useless is that we have ignored the **direction** of the ball's motion. The average speed is thus not a significant quantity. That is why, for the physical sciences, **velocity is commonly used, rather than speed.**

To sum up the quantitative description of motion:

1.   The quantities displacement, velocity, acceleration, and time are used.

$$(1) \quad a_X = \frac{\Delta v_X}{\Delta t} \qquad \text{and} \qquad (2) \quad a_Y = \frac{\Delta v_Y}{\Delta t}$$

2.   The basic equations (assuming that the acceleration is constant) are

$$(3) \quad \bar{v}_X = \frac{\Delta X}{\Delta t} \qquad \text{and} \qquad (4) \quad \bar{v}_Y = \frac{\Delta Y}{\Delta t}$$

$$(5) \quad \bar{v}_X = \frac{1}{2}(v_{1X} + v_{2X}) \qquad \text{and} \qquad (6) \quad \bar{v}_Y = \frac{1}{2}(v_{1Y} + v_{2Y})$$

3.   For an object moving along a straight line, we may use either the three X equations or the three Y equations. In case the object is not moving on a straight line, for example, a thrown or batted baseball, we use both the X and the Y equations.

EXAMPLE 1.25

Consider a 3000-lb car that is moving at 30 mi/h. It hits a concrete barrier and is stopped. Measurement shows that the front of the car was pushed in 1 ft by the collision. This indicates that the car moved forward 1 ft as it was being brought to a stop. Determine the following:

1. The average velocity of the car as it was coming to a stop
2. The acceleration of the car as it was coming to a stop
3. The amount of force that acted on the car as it was coming to a stop

Solution

This example involves motion in one direction, so we need to use only the X or the Y equation. Since Y usually refers to vertical motion, we will use X.

Before we write the given information there are three decisions to be made:

1. We will pick forward to be the $+X$ direction.
2. We must identify the two instants 1 and 2. Let instant 1 be when the car just hits the barrier and instant 2 be after the car has come to a stop.
3. We must pick a system of units. We will use the USA system in this example.

$$W_{car} = 3000 \text{ lb}$$
$$v_{1X} = 30 \text{ miles per hour}$$
$$v_{2X} = 0 \text{ (because the car has come to a stop)}$$
$$\Delta X = 1 \text{ ft}$$
$$t_1 = 0$$

The weight of the car is given in pounds; this is okay. The initial velocity is given in miles per hour. This must be changed to feet per second:

$$\left(30 \frac{\text{miles}}{\text{hour}}\right)\left(\frac{1 \text{ hour}}{3600 \text{ seconds}}\right)\left(\frac{3.281 \text{ feet}}{6.214 \times 10^{-4} \text{ miles}}\right) = 44 \frac{\text{ft}}{\text{s}}$$

The distance is given in feet, this is okay.
The quantities to be calculated are as follows:

    a. Average velocity: $\bar{v}_X$
    b. Acceleration: a_X
    c. Force: F_X

Calculations:

    a. The average velocity is given by

$$\bar{v}_X = \frac{1}{2}(v_{1X} + v_{2X})$$

    b. The formula for acceleration is

$$a_X = \frac{\Delta v_X}{\Delta t}$$

$$a_X = \frac{v_{2X} - v_{1X}}{t_2 - t_1}$$

$$a_X = \frac{0 - 44}{t_2 - 0}$$

$$= \frac{-44}{t_2}$$

To calculate a_X, we first must determine t_2. Consider the formula for the displacement:

$$\Delta X = \bar{v}_X \, \Delta t$$

We know that the displacement is 1 ft, and we have already determined the average velocity, so

$$1 = (22)(t_2 - 0)$$
$$t_2 = 0.045 \text{ s}$$

This shows that it takes 0.045 s (= 45 ms) for the car to come to a stop. This seems such a very small amount of time, but it is a very reasonable Δt for an automobile collision.

We can now determine the acceleration:

$$a_x = \frac{-44}{t_2}$$

$$a_x = \frac{-44}{0.045}$$

$$= -977 \frac{\text{ft}}{\text{s}^2}$$

This may be expressed in terms of g's:

$$a_x = -31 \text{ g's}$$

c. We can now determine the amount of force acting on the car:

$$F_x = (W)(a_{g's})$$
$$= (3000 \text{ lb})(-31)$$
$$= -93 \times 10^3 \text{ lb}$$

Notice the negative sign. Since we chose forward to be the positive direction, the negative signs mean that the acceleration points backward and also that the force, that is stopping the car, points backward.

Such a large amount of force will certainly explain the great amount of damage done to the front of the car. We will consider what happens to the driver and passenger later, in the section on momentum and impulse.

EXAMPLE 1.26

We can use these equations to gain some appreciation for the difficulty involved in hitting a fastball. The following information is given:

1. Distance from the pitcher's mound to home plate: 18 m
2. Initial velocity of the pitched ball: 90 mi/h = 40 m/s
3. Time necessary for bat to move from its original position to the plate: 200 ms

Analysis:

We will pick the $+X$ direction from the pitcher to the batter. Since we ignore air resistance, the only force that acts on the ball as it moves toward the batter is gravity. Therefore, $a_x = 0$. Equation (1) implies that $v_{1x} = v_{2x} = 40$ m/s. Then equation (5) implies that the average velocity equals 40 m/s. We can then use equation (3) to determine the amount of time that the ball takes to travel from the pitcher to batter:

$$\bar{v}_X = \frac{\Delta X}{\Delta t}$$

$$40 = \frac{18}{\Delta t_{ball}}$$

$$\Delta t_{ball} = 0.45 \text{ s}$$

The ball takes 450 ms to travel to the plate, and we are given that 200 ms are required for the swung bat to reach the plate. This means that the batter has 250 ms after the ball is thrown to decide whether or not to swing and, in general, where (high, low, inside, or outside) to expect the ball. During this 250 ms, the ball will travel 10 m; it is still 8 m (24 ft) away from the batter. It may still curve in or out. If the batter could get the bat around in less time than 200 ms, there would be more time to decide about the swing. **This is why the use of a lighter bat often leads to higher batting averages.**

EXAMPLE 1.27

A reasonable value for the initial velocity of a hard hit ball is 120 mi/h at an angle of 35° above the horizontal. Assume that such a ball is hit at a point that is 40 cm above home plate.

    1. How far (assuming that nothing gets in the way) will it travel before it hits the ground?
    2. How much time will it spend in the air?

Solution:

    1. Decide on a coordinate system: Let the origin be at home plate, $+X$ pointing away from batter, and $+Y$ pointing up.
    2. Identify the two instants 1 and 2: Instant 1 is when ball is hit, and instant 2 is when it hits the ground.
    3. Pick a system of units, and write the given information in that system. Use the SI system:

$v_1 = 120$ mi/h $= 53.6$ m/s
$v_{1X} = 53.6 \cos 35 = 43.9$ m/s $\qquad\qquad v_{1Y} = 53.6 \sin 35 = 30.8$ m/s
$X_1 = 0$ $\qquad\qquad\qquad\qquad\qquad\qquad\quad Y_1 = 0.4$
$a_X = 0$ (assuming that we ignore air resistance) $\qquad a_Y = -9.8$ m/s^2
$Y_2 = 0$ (ball hits the ground) $\qquad\qquad\qquad t_1 = 0$
$X_2 = ?$ $\qquad\qquad\qquad\qquad\qquad\qquad\quad t_2 = ?$

    4. Write the basic equations (*Note*: Since the problem involves two dimensions, X and Y, we use all six equations):

$$(1) \quad a_X = \frac{\Delta v_X}{\Delta t} \qquad (2) \quad a_Y = \frac{\Delta v_Y}{\Delta t}$$

$$(3) \quad \bar{v}_X = \frac{\Delta X}{\Delta t} \qquad (4) \quad \bar{v}_Y = \frac{\Delta v}{\Delta t}$$

$$(5) \quad \bar{v}_X = \frac{1}{2}(v_{1X} + v_{2X}) \qquad (6) \quad \bar{v}_Y = \frac{1}{2}(v_{1Y} + v_{2Y})$$

5. Substitute the given information into the basic equations:

(1) $0 = \dfrac{v_{2x} - 43.9}{t_2 - 0} \Rightarrow v_{2x} = 43.9$

(2) $\bar{v}_x = \dfrac{43.9 + 43.9}{2} = 43.9$

(3) $43.9 = \dfrac{x_2 - 0}{t_2 - 0} = \dfrac{x_2}{t_2}$

(4) $-9.8 = \dfrac{v_{2y} - 30.8}{t_2 - 0}$

(5) $\bar{v}_y = \dfrac{v_{2y} + 30.8}{2}$

(6) $\bar{v}_y = \dfrac{0 - 0.4}{t_2 - 0}$

6. Use algebra to combine the equations:
   a. Combine (6) and (4):

$$\dfrac{-0.4}{t_2} = \dfrac{v_{2y} + 30.8}{2} \Rightarrow v_{2y} = \dfrac{-0.8}{t_2} - 30.8$$

   b. Substitute into (2):

$$-9.8 = \dfrac{\left(\dfrac{-0.8}{t_2} - 30.8\right) - 30.8}{t_2}$$

$$-9.8t_2 = \dfrac{-0.8}{t_2} - 61.6$$

$$-9.8t_2^2 + 61.6t_2 + 0.8 = 0$$

use the quadratic formula:

$$t_2 = \dfrac{-61.6 \pm \sqrt{(61.6^2 - 4(-9.8)(0.8))}}{2(-9.8)}$$

$$= -0.013, 6.3$$

Since an instant of time, such as t_2, cannot be negative, $t_2 = 6.3$ s
   c. Substitute into (3):

$$43.9 = \dfrac{x_2}{6.3} \Rightarrow 276.5 \text{ m } (= 907 \text{ ft})$$

   This answer, while it is mathematically correct, is physically impossible. The distance comes out impossibly long (the fence of a baseball park is less than 500 ft from home plate) because we neglected air resistance. Although the omission leads to an unrealistic result, including air resistance would lead to the need for very involved mathematics. This calculation is intended to demonstrate the equations, not to arrive at a realistic answer.

PROBLEM SET 2

2.1. Three masses hanging from a force table.

	ANGLE	MAGNITUDE (GRAMS)
F_1	50°	200
F_2	175°	300
F_3	260°	100

Where should a fourth pulley be placed, and how many grams should be hung from it so that the central ring will be in equilibrium? (204 grams at 337°.)

2.2. A 20-lb weight is suspended from two cords as shown in the diagram. The cord on the right exerts a force of 15 lb at an angle of 30° with the vertical.
   a. How much force is exerted by the cord on the left? (10.3 lb)
   b. Determine the size of the angle between the left cord and the vertical. (47°)

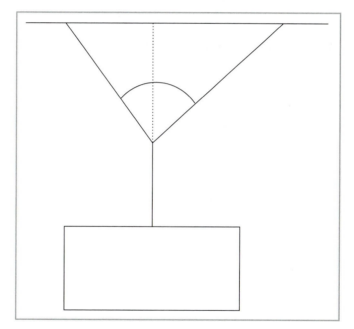

2.3. The following figures represent a person's right hip. Figures (a) and (b) represent a normal hip, figure (c) represents what is called *coxa valga,* and figure (d) represents what is called *coxa vara.* In the normal case, the angle of inclination (between the axis of the shaft of the femur and the neck of the femur) is 125° Coxa valga represents a larger than normal angle of inclination, and coxa vara represents a smaller than normal angle. The force F_1 represents the amount of weight that is pushing vertically down on the femur-acetabulum joint. Assume that this force is 70 pounds and that the angle between the vertical and the shaft of the femur is 5°. Consider the normal case (figure b), and answer the following questions:
   a. How much force is acting along the axis of the neck of the femur? (*Note:* This force is compressing the joint.) (45 lb)

b. How much force is acting perpendicular to the axis of the neck of the femur?

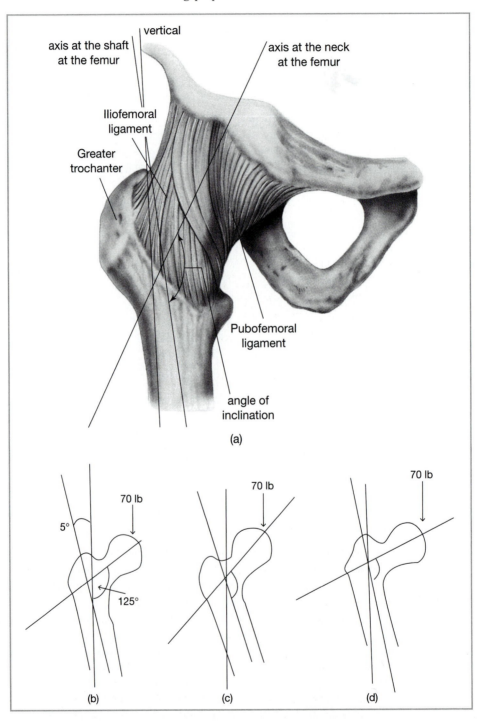

vertical

axis at the shaft
at the femur

axis at the neck
at the femur

Iliofemoral
ligament

Greater
trochanter

Pubofemoral
ligament

angle of
inclination

(a)

70 lb

5°

125°

(b)

70 lb

(c)

70 lb

(d)

(*Note:* This force is causing slippage at the joint.) (53.6 lb)

   c. Which of the three cases (figures (b), (c), or (d)) represents the most stable situation? Explain your answer.

2.4. In the diagram, Fe represents the force of the erector spinae muscle, and W represents the weight (90 lb) of the trunk of the body. There is also a line (with no arrows) that represents the axis of the spinal column. The person is bending over so that the spinal column makes an angle of 30° with the horizontal. The erector spinae muscle makes an angle of 12° with the axis of the spinal column.

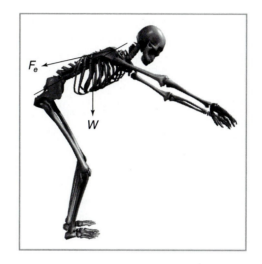

   a. Assume that it is necessary that the pull of the erector spinae muscle results in a force of 55 N that will cause the back (spinal column) to rotate. Determine the tension (or force) produced by the erector spinae muscle. (265 N)

   b. How much force is causing compression of the spinal column? Limit your analysis to the two forces shown in the figure. (304 N)

2.5. Explain why the turns at a high-speed raceway are usually banked.

2.6. How is giving directions in a city related to components?

2.7. A car is moving at a constant velocity of 45 miles per hour. It has been experimentally determined that it takes a typical driver 250 ms to apply the brakes, once an impending collision is recognized. How far will the car move during this interval of time? (16.5 ft)

2.8. A car is moving at 45 miles per hour when the driver sees a tree down across the road. Assume that the maximum acceleration associated with the brakes is −1.25 g's. The driver's reaction time (time taken to hit the brakes) is 250 ms.

   a. How far does the car travel before the driver applies the brakes? (16.5 ft)

   b. How far does the car travel while it is slowing down? (54.5 ft)

   c. Assume that the car's headlights are set to illuminate objects that are up to 120 ft in front of the car. What is the highest velocity at which the car may be safely driven at night? (60 miles/hour)

NEWTON'S SECOND LAW

We have been using the idea, based on the work of Galileo, that if there is no external force acting on an object, then that object's velocity will remain constant and hence its acceleration will equal zero. That is, if it happens to be at rest, it will remain at rest. If it happens to be moving, it will continue to move at constant speed in a straight line. This idea was later restated by Isaac Newton and is known as Newton's first law. We have also shown that if there are several forces acting on an object such that the forces happen to add up to zero, then again the velocity will remain constant. Now we shall deal with situations in which there are several forces acting, that do not add up to zero.

This analysis was developed by Newton and is named Newton's second law in his honor. According to this idea, if there is a nonzero net (total) force acting on an object, then it will result in acceleration. The relation between the net force, and the resulting acceleration is

$$\vec{F}_{net} = m\vec{a}$$

Since this is a vector equation, it tells us that the direction of the net force and the direction of the acceleration must be the same. This will be very important when we discuss centripetal force later.

The left-hand side of the equation is the net force acting on the object. If there is only one force acting, then that one force is the net force. If there are several forces acting on the object, then the net force must be determined by adding the several forces (keeping in mind that they are vectors).

The adding process is indicated by the Greek capital sigma

$$\sum \vec{F}_{external} = m\vec{a}$$

where $\sum \vec{F}_{external}$ represents the sum or resultant of the external forces acting on the object, m is the mass, and $\vec{a}$ is the acceleration.

Notice that this vector equation is very similar to the summation of the forces equation that we used in the previous section. The left-hand side is, once again, the sum of the external forces. However, the right-hand side is not zero as it was before; it is the product of the mass of the object multiplied by its acceleration. The procedure for using this equation will be almost identical to the procedure discussed in the previous section.

Once again we will use the scalar equations that together represent the vector equation:

$$+ \rightarrow \sum F_X = ma_X$$

$$+ \uparrow \sum F_Y = ma_Y$$

(Note: The arrows indicate the positive X and Y directions, respectively.)

It is very important to note that the terms that appear on the left-hand side of the equations represent, the X and Y components of physically real forces that act on the object.

Examples of such forces are the following:

1. The **force of gravity,** also called the **weight** of the object. This force always acts vertically down toward the center of the Earth.
2. The force exerted by a muscle, a tendon, a cable, a string, or the like. This type of force always pulls on the object, never pushes, and is often called **tension.**
3. The **force of friction** between the object and a surface. This force is always parallel to the surface and points opposite to the direction in which the object is sliding or tending to slide.
4. The **normal or contact force** between an object and a surface. This force is always directed perpendicular to the surface and points toward the object. The word "normal" is not used in the sense of contrast with abnormal. It is used to mean perpendicular.

Sometimes, the force that a surface exerts on an object is not given in a problem. Rather, the force exerted by the object on the surface is given or is easily calculable. It is very straightforward to determine the magnitude and direction of the force that object A exerts on object B if the force that B exerts on A is known. This is accomplished by the use of Newton's third law (discussed earlier).

Returning to the second law, the terms on the right-hand side, ma_X and ma_Y, do not represent forces and should not initially be included on the left-hand side in writing the basic equations. They represent the components of the acceleration of the object rather than external forces. However, after writing the basic equations, algebra may be used in later steps to move terms from one side to the other.

Newton's second law is one of the basic relations in classical physics. It means that if an object is speeding up, some force originating outside of the object must be making it do so. The same is true if the object is slowing down. So when the dropped rock is speeding up, the force of gravity is the cause. Or when a thrown rock slows as it rises, the force of gravity is the cause. When a speeding car slows to a halt, it is the force of friction of the road on the tires (an external force) that is the cause, not the friction within the brakes of the car (an internal force). If an object changes direction, there must be an external force causing that change in direction. In the case of a planet orbiting a star, it is the force of gravity that causes the planet to change direction so that it will follow a closed orbit.

EXAMPLE 1.28

One of the body's functions that are most subjected to research is **gait.** This is the collection of muscle, tendon, and bone interactions that result in walking and running. There are several models, or representations, of a person walking that are used in gait analysis. The following discussion describes the most basic model.

Assume that a woman is walking such that:

1. at no time are both feet off the ground and
2. each leg is kept straight while its foot is on the ground.

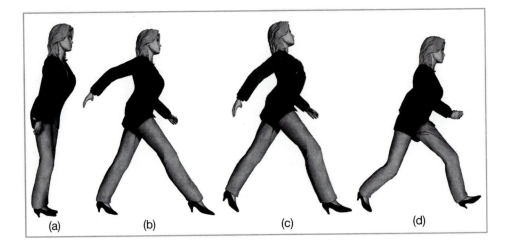

(a)   (b)   (c)   (d)

We see that in figure (a), her left foot has just hit the ground, thus coming to a stop. Her right foot has just left the ground and is moving forward, speeding up as it goes. In figure (b), both legs are vertical, and the right foot is at midstride. In figure (c), the right foot is still moving forward but is slowing down, preparatory to making contact with the ground. The left foot is still in contact with the ground and will soon be lifted up. In figure (d), the right foot has made contact with the ground and is not moving. The left foot has left the ground and is speeding up.

Between figures (a) and (b), the right leg is speeding up. An external force is required to cause this acceleration. That force is supplied by several muscles that connect the front of the hip (ilium) to the front of the leg (femur and tibia). When these muscles contract, they accelerate the leg forward. This acceleration (speeding up) continues until figure (c), in which, the leg is moving at its maximum speed. Between figures (c) and (d), the right leg is slowing down. This is caused by the gluteus maximus muscles that connect the rear of the ilium to the rear of the femur. When this muscle contracts, it pulls the leg backward, slowing it down. In figure (d), the right leg has come to a stop.

EXAMPLE 1.29

We have already discussed Newton's Law of Gravitation and the accepted value of the constant, G. It is now possible to derive an expression for the acceleration of an object that is in free fall near the Earth's surface. (*Note*: This had already been experimentally determined to be 9.8 m/s^2.) Consider an object of mass m that is in free fall (acted upon only by the force of gravity) near the surface of the Earth. We can use Newton's second law to determine its acceleration:

$$\vec{\mathbf{F}}_{net} = m\vec{\mathbf{a}}$$

The force on the left side is given by

$$F_{gravity} = G \frac{M_{earth} m_{object}}{r_{earth}^2}$$

$$G \frac{M_{earth} m_{object}}{r_{earth}^2} = m_{object} a_{object}$$

$$a_{object} = G \frac{M_{earth}}{r_{earth}^2}$$

We have already agreed that the acceleration of an object, due to gravity alone, will be designated as g, the acceleration of gravity:

$$g = G \frac{M_{earth}}{r_{earth}^2}$$

Notice that g depends only on the distance between the object and the center of the earth. So it is reasonable that all objects that are near the surface of the earth, affected only by gravity, will experience the same acceleration. Of course, if an object happens to be significantly farther from the center of the earth than its surface, the acceleration will be different. If we substitute the accepted value of the acceleration of gravity at the surface of the earth, the value of G, and of the radius of the earth, we can calculate the mass of the earth:

$$M_{earth} = \frac{g_{surface} r_{earth}^2}{G}$$

Substituting SI values for the quantities on the right-hand side, we have

$$M_{earth} = \frac{\left(9.8 \frac{m}{s^2}\right)(6.37 \times 10^6 \, m)^2}{(6.673 \times 10^{-1} \, SI)}$$

$$M_{earth} = 6 \times 10^{24} \, kg$$

The procedure for solving Second Law problems is as follows:

1. Identify the object of interest.
2. List the forces that act **on** that object.
3. Make a free body diagram
4. Write the scalar equivalent equations. These will always be the same:

$$+ \rightarrow \sum F_X = ma_X$$

$$+ \uparrow \sum F_Y = ma_Y$$

5. Determine the X and Y components of each known vector.
6. Enter all of these components into the scalar equations.
7. Solve for the unknown components.
8. Make a sketch showing them on a set of axes.

9. Draw the rectangle bounded by the components.
10. Draw the vector pointing from the origin to the opposite corner of the rectangle.
11. Use trigonometry to solve for the magnitude and direction of that vector.

EXAMPLE 1.30

In the preceding section, we discussed a woman standing on a bathroom scale. For some reason, she took this scale with her into elevators and worried about the readings. We can now analyze what was going on. In particular, we will see why she seemed to be putting on weight and taking off weight.

Solution

1. Let us assume that the object of interest is the woman's body. After all, this is what the forces (gravity and the scale) are acting on.
2. The forces acting on her are
   a. gravity, pulling down: F_g
   b. the scale, pushing up: F_s
3. The FBD is shown on the right:
4. The scalar equation is

$$+ \uparrow \sum F_y = ma_y$$

Steps 5–7 are done as follows:

$$F_s - F_g = ma_y$$
$$F_s = F_g + ma_y$$

If we associate her weight (W) with the force of gravity (F_g),

$$F_s = W + ma_y$$

Now we can analyze the scale readings as compared to her weight. **If the elevator were going up and then slowed down to a stop, its acceleration would point down (thus a_y would be negative). In this case, the scale reading would be less than her weight. If the elevator were going down and slowed to a stop, its acceleration would point upward (thus a_y would be a positive number). In this case, the scale reading would be more than her weight.**

EXAMPLE 1.31

A 200-lb boat is acted on by two horizontal forces: one from the wind and the other from the current in the water. The wind is represented by a force of 150 lb directed forward and to the right at 15° away from the heading of the boat. The current is represented by a force of 75 lb directed perpendicular to the heading of the boat, to the right. Determine the magnitude and direction of the acceleration of the boat. (*Note*: The vertical forces—the force of gravity and the buoyancy of the water—will cancel and therefore play no part in this problem.)

Solution

1. The problem states that the forces act on the boat, so the boat is the object of interest.
2. The forces acting on the boat are as follows:
   a. F_{wind}: 150 lb, 15° to the right of the heading of the boat
   b. $F_{current}$: 75 lb, to the right, perpendicular to the heading
   c. $F_{gravity}$ may be ignored.
   d. $F_{buoyant}$ may be ignored.
3. The figure shows a free body diagram of the boat, viewed from directly above. The $+X$ axis has been chosen to represent the direction of the heading of the boat.

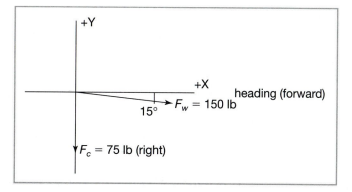

4. The basic equations would be

$$+ \rightarrow \sum F_X = ma_X$$
$$+\uparrow \sum F_Y = ma_Y$$

5. Substituting from the FBD into the two equations, we get:

$$150 \cos 345° + 75 \cos 270° = (200/32)(a_X)$$
$$150 \sin 345° + 75 \sin 270° = (200/32)(a_Y)$$

6. Solving these equations: we have

$$a_X = 23.18 \text{ ft/s}^2$$
$$a_Y = -18.21 \text{ ft/s}^2$$

7. Since a_X comes out positive, it points in the $+X$ direction (forward), and since a_Y comes out negative, it points in the $-Y$ direction (to the right of forward).

8.

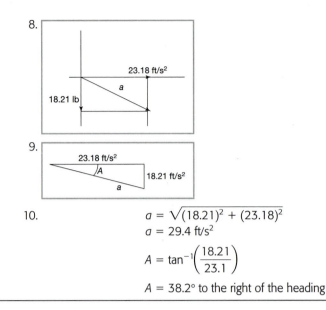

9.

10.

$$a = \sqrt{(18.21)^2 + (23.18)^2}$$
$$a = 29.4 \text{ ft/s}^2$$

$$A = \tan^{-1}\left(\frac{18.21}{23.1}\right)$$

$$A = 38.2° \text{ to the right of the heading}$$

EXAMPLE 1.32

The figure shows a boy pulling a 10-kg box along a horizontal floor. There is 1.5 N of frictional force between the box and the floor. The angle between the rope and the horizontal is 30°. The box is observed to accelerate to the left at 0.5 m/s².

1. Determine the tension in the rope.
2. Determine the coefficient of kinetic friction between the box and the floor.

Solution:

1. Since the box is the object that is accelerating and is also what the rope and friction are acting on, it is reasonable to assume that the box is the object of interest.
2. The forces acting on the box are as follows:
   a. Rope, F_r, magnitude unknown, directed up and to the left at 30° above the horizontal.
   b. Gravity, the weight of the box, W, magnitude not given but known to be equal to mg, directed straight down (because all weights are directed straight down)
   c. Friction, F_f, magnitude given as 1.5 N, directed to the right (because the box is sliding to the left)
   d. Normal force, F_n, magnitude not given but known to be related to F_f by $F_f = \mu F_n$, directed vertically up (because a normal force is always perpendicular to the surface and points toward the object)
3. The free body diagram is as follows:

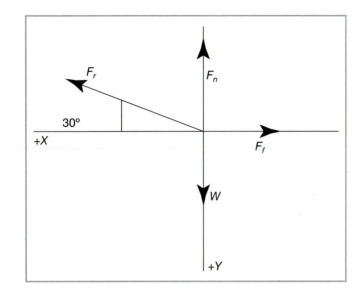

4. Notice that the $+X$ direction has been chosen to be to the left because that is the direction given for the observed acceleration. The $+Y$ direction is down because it must be 90° CCW from $+X$.

5. The basic equations are

$$+ \to \sum F_X = ma_X$$

$$+ \uparrow \sum F_Y = ma_Y$$

and, since friction is involved,

$$F_f = \mu F_n$$

6. Substituting from the FBD into the basic equations, we have

$$F_r \cos 30° - F_f = ma_X$$
$$W - F_r \sin 30° - F_n = ma_Y$$

7. We are given in the statement of the problem that

$$a_X = 0.5 \text{ m/s}^2$$
$$a_Y = 0$$
$$m = 10 \text{ kg}$$
$$F_f = 1.5 \text{ N}$$

Substituting these values into the three equations, we get:

$$1.5 = \mu F_n$$
$$F_r \cos 30° - 1.5 = (10)(0.5)$$
$$(10)(9.8) - 1.5 \sin 30° - F_n = (10)(0)$$

8. We now have three equations involving three unknown variables, which may be evaluated:

$$F_r = 7.5 \text{ N}$$
$$F_n = 97.25 \text{ N}$$
$$\mu = 0.08$$

CENTRIPETAL FORCE

We have been dealing with situations in which an object speeds up or slows down, and we have shown that there is always an external force that causes such changes. We will now consider an object that is not speeding up or slowing down but is changing direction.

The diagram shows a highway curve and the front seat of a car. There are two people in the car, the driver (D) and the passenger (P). Notice that in the first two positons, the car is on a straight section of road; P and D are sitting close to each other. They are obviously friendly but, knowing the rules of the road, are not holding onto each other. We will assume that the two people are not wearing seat belts and that the car seat is so smooth that there is no friction between them and the seat.

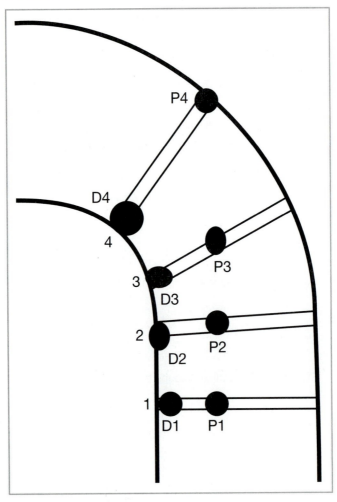

As the car moves around the curve, we have position 3. At this point, there is the following conversation:

D: "Why are you moving away from me?"

P: "I'm not trying to move, something is making me slide."

D: "Well, all right, but I don't think much of it."

A critical point is reached in position 4. Either the door is opened and P and D part company or the door remains closed and the rest of the trip goes on in silence.

Consider lines drawn showing the paths of D and of P. **Notice that P is traveling on a straight line.** P was not being held by D, there was no friction between P and the car seat, the force of gravity was pulling P down, but the car seat was pushing P up. Therefore the net external force acting on P is zero, and the resulting path is a straight line. This is an example of Newton's first law.

Notice that D's path is a curved line. Therefore it is not P who is moving away from D, but D who is moving away from P. What about the external forces that act on D? Once again, the force of gravity pulls down, and the car seat pushes up (these two forces canceling each other); there is no friction between D and the car seat, but there is another interaction (force). D is holding onto the steering wheel. As the car moves around the curve, D is aware of pulling out on the steering wheel. From knowledge of Newton's third law, we know that the steering wheel must be pulling in on D. Thus there is a net external force on D, and it points toward the center of the curve (or arc).

This is an example of a general statement:

Whenever an object changes direction, there is a net force acting on that object, and that force points toward the center of curvature of the arc.

We can approach this situation more mathematically. Consider D's path. We can analyze D's acceleration from the definition:

$$\vec{a} = \frac{\Delta \vec{v}}{\Delta t}$$

Consider two points on D's path: D2 and D3. We can determine the direction of $\Delta \vec{v}$. Recalling the definition of the symbol Δ, we have :

$$\Delta \vec{v} = \vec{v}_3 - \vec{v}_2$$
$$\Delta \vec{v} = \vec{v}_3 + (-\vec{v}_2)$$

Representing this equation graphically, we see that $\Delta \vec{v}$ points down and to the left, that is, toward the center of the curve. Since $\vec{a}$ points in the same direction as $\Delta \vec{v}$, it must also point toward the center of the curve and from Newton's second law ($\vec{F}_{net} = m\vec{a}$), the net force must also point toward the center of the curve.

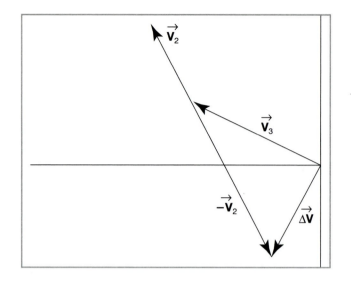

Any acceleration that is related to a change in direction, not to a change in speed, must point toward the center of the curve and is called **centripetal acceleration,** a_c. A more careful analysis yields the magnitude as well as the direction of the centripetal acceleration.

The figure shows an object moving around a circle (radius $= r$) at constant speed (v). We know that the object will experience an acceleration (a_c) directed inward, toward the center of the circle.

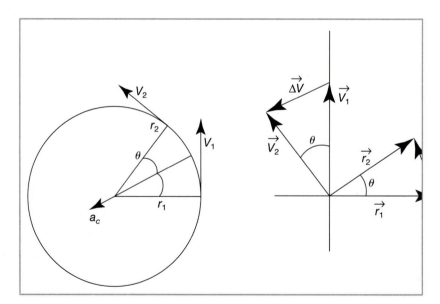

We see from the figure on the right that, since $r_1 \perp v_1$ and $r_2 \perp v_2$, the two angles labeled θ are equal. The two triangles are both isosceles, and since they have a common apex angle, they are similar:

$$\frac{\Delta v}{v} = \frac{\Delta r}{r}$$

$$\Delta v = \frac{v}{r}\,\Delta r$$

Dividing both sides by Δt, we have

$$\frac{\Delta v}{\Delta t} = \frac{v}{r}\frac{\Delta r}{\Delta t}$$

but since

$$\frac{\Delta r}{\Delta t} = v \qquad \text{and} \qquad \frac{\Delta v}{\Delta t} = a$$

we have

$$a = \frac{v^2}{r}$$

So we have

$$a_{\text{centripetal}} = \frac{v^2}{r}$$

Substituting this acceleration into Newton's second law, we see that

$$F_{\text{centripetal}} = \frac{mv^2}{r}$$

directed radially inward toward the center of the arc.

This equation defines **centripetal force**. It comes into play whenever an object changes direction. There are many physical forces in the world that may play the role of a centripetal force. When a planet goes around a star, it is changing direction; There must be a centripetal force, and there is: the force of gravity of the star on the planet. When a car goes around a corner on a level road, it is changing direction; there must be a centripetal force, and there is: the force of friction of the road on the tires of the car. Similarly, when an electron goes around a nucleus in an atom, it is the electrical force of attraction of the nucleus of the atom on the electron that plays the role of the centripetal force. Whenever anything changes direction, there must be a physical force that plays the role of a centripetal force.

The figure shows what happens when the force of friction between the road and the tires is not large enough to provide the required centripetal force. The car is moving clockwise around the track. If the force of friction is large enough, the car will negotiate the turn, and all will be well. If the force of friction is not large enough, the car will follow the path shown by the straight-line skid marks and go off the track.

It is important to note that centripetal force is not an additional physical force. When the second law is written, centripetal force cannot appear on the left-hand side with the physical forces. It must appear on the right-hand side, since it is fundamentally based on acceleration.

$$\sum \vec{F}_{\text{external}} = m\vec{a}$$

$$\sum \vec{F}_{\text{external}} = m\frac{v^2}{r}, \quad \text{radially inward}$$

The force(s) on the left side of the equation must be physical forces, such as friction, gravity, electric, or magnetic.

EXAMPLE 1.33

Consider a roller coaster ride in an amusement park. The ride is constructed so that the car moves only in a vertical plane. That is, the car goes only up and down as it proceeds along the track; it never moves to the left or to the right. At the end of the ride, one of the passengers complains of neck pain, and it is later determined that there was compression damage to the cervical vertebrae.

1. At what point during the ride did this injury most probably happen?
2. Referring to the physical concepts used in this course, explain why the injury happened.

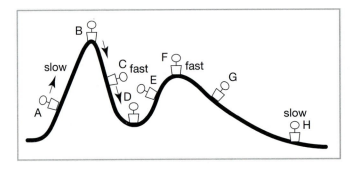

Solution

1. The injury probably happened at point D in the figure.
2. At point D, the car is changing direction while moving at high speed around a tight arc (one with a small radius of curvature). Since it is changing direction, it and the people in it experience a centripetal acceleration and are therefore subject to a centripetal force. This force, by definition, points toward the center of the arc—in this case, straight up. The centripetal force acting on the rider's head would be represented by the cervical vertebrae pushing up on the bottom of the skull. From Newton's third law, we know that the skull will push down on the vertebrae. **Therefore the vertebrae will be under compression and may fracture.** (*Note*: Centripetal force is also playing a role at point B. However, at this point, the center of the arc is below the track, so the net force on the person's head will point down. Therefore, the head is pulling up on the vertebrae

(Newton's third law) and thus they—actually, the tendons and other soft tissue—will be in tension, not compression.)

EXAMPLE 1.34

Imagine a person who is acted on by two forces: the force of gravity pulling down and some other force pushing (or pulling) up. If the second force is represented by a scale (e.g., a bathroom scale), we usually interpret its reading as the person's weight. In fact, we commonly associate the weight of any object with the reading of a scale that is supporting it. Let's see how this may be described by using the second law.

Consider the FBD to the right.

Applying the second law, we have

$$\sum \vec{F}_{\text{external}} = m\vec{a}$$

and picking down to be the positive direction, we get

$$F_{\text{gravity}} - F_{\text{scale}} = ma$$
$$mg - F_{\text{scale}} = ma$$

Case 1: The person is at rest: Consider a 120-lb person who is sitting in a chair in a classroom in Hartford, Connecticut (latitude 41.77°).

1. Is this person in equilibrium?
2. Find the magnitude of the net force that acts on the person.

Solution

a. No, she is not in equilibrium. She is accelerating. We know this because she, like everything else in Hartford, is moving around a circle of radius 2968 miles every 24 hours. Therefore the direction of her velocity is changing; she is accelerating.

b. Since she is accelerating, we know that she is experiencing a force, in particular a centripetal force (i.e., there is a nonzero net force pointing down). We may calculate its magnitude from the definition:

$$F_{\text{centripetal}} = \frac{mv^2}{r}$$

She moves around a circle of radius 2968 miles in 24 hours. Therefore her speed is 777 miles per hour (1139.6 ft/s). This results in a force of 0.3 lb. This is a very small force—0.25% of her weight, one that a person would be totally unaware of. Nevertheless, it is not zero.

We see that even though her speed is very large (777 miles per hour), the radius associated with the path is so large (2968 miles) that the resulting acceleration (and force) is very small. For most real applications, the amount of centripetal acceleration or centripetal force that is due to the rotation of the earth is so small that it may be ignored. However, if the path had a much smaller radius, a noticeably larger effect might result.

Case 2: The person is moving. Consider the same person on the same scale, but this time in a bus that is moving on a road. If the road is flat, then the radius of the path is once again the radius of the Earth, and the associated centripetal effects are negligible. But suppose that the bus goes up and over a hill.

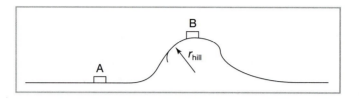

Examine the situation at point B in the figure. Gravity pulls the person down (toward the center of the Earth), and the supporting scale pushes the person up. The motion of the bus up and over the hill requires a changing direction, implying centripetal force. If we take down (toward the center of the Earth) as positive, the second law would be

$$F_{gravity} - F_{scale} = m_{person}\frac{v^2}{r_{hill}}$$

$$m_{person}g - F_{scale} = m_{person}\frac{v^2}{r_{hill}}$$

The reading on the scale, which we usually interpret as the person's weight, is given by

$$F_{scale} = m_{person}g - m_{person}\frac{v^2}{r_{hill}}$$

If this trip over the hill were to be tried several times, each faster than the one before, the equation predicts that at a critical speed,

$$v_{critical} = \sqrt{r_{hill}\,g}$$

The reading on the scale would be zero. The person would appear to be weightless. (*Note*: Review material on pages 20–21). When this really happens (moving over the top of a vertical curve), you actually feel as though you are floating off your seat. If this happens in a roller coaster, for example, you may have to depend on a safety bar or seat belt to keep you in the car.

Notice that, at this critical velocity, the person's acceleration is equal to the acceleration of gravity at that point. This is called **free fall**. The person is not falling but is experiencing the acceleration of gravity and, as a result, experiencing **weightlessness**. This may be generalized as follows:

If the acceleration of an object is equal (in both magnitude and direction) to the acceleration of gravity, it will appear to the object that the force of gravity is zero.

This may be accomplished in several ways:

1. A person actually falling (assuming that air resistance may be neglected)

2. A person on a roller coaster, airplane, bus, or some other vehicle that is moving over the top of a vertical curved path at the critical velocity for that particular path

3. A person in a space vehicle that is in orbit at the critical velocity for the radius of that orbit

EXAMPLE 1.35

Refer to the figure on page 60, which shows a sequence of pictures of a woman (mass M) walking at a constant speed (v). Imagine an origin located at her hips, and focus attention on one of her feet. The foot would move from being located forward of the hip to below the hip, then behind the hip, again below the hip, and then forward of the hip. This process is repeated over and over. Thus it seems as though the foot is moving back and forth on the arc of a circle. Now imagine that the origin is located on the foot, and focus attention on the hip. The hip moves from a location behind the foot to being above the hip, then forward of the hip, above the hip, and then behind it. Thus from the point of view of the foot, the hip seems to be moving back and forth on the arc of a circle. This situation may be modeled by representing the body by a point at the center of gravity, moving on the arc of radius L (the length of her legs).

Since the body is moving on an arc whose center is below the arc, the centripetal force must point vertically down. Her body will experience a centripetal force given by:

$$F_{centripetal} = \frac{Mv^2}{L}, \quad \text{down}$$

The only real force that could play this role is the force of gravity, her weight (Mg), and therefore the centripetal force cannot be greater than her weight. Thus the second law becomes

$$Mg \geq \frac{Mv^2}{L}$$

$$v \leq \sqrt{gL}$$

Assuming that her legs are 0.8 m long, the maximum speed at which she could walk would be about 2.8 m/s. If she wanted to move faster than this, she would automatically start to run. People who engage in racewalking (an Olympic sport) must learn to walk in a manner that avoids the limitation on speed that is described here. If you have ever seen such races, it is evident that the competitors are not walking naturally.

This calculation compares favorably with experiments that show that an adult automatically breaks into running when trying to walk faster than about 2.5 m/s (about 5.6 mi/h). The equation also explains why a child must run to keep up with a quickly walking adult.

PROBLEMS SET 3

3.1. A pitcher throws a regulation 115-gram (0.253-lb) baseball at a speed of 100 mi/h. From the end of the windup to when the ball was released, 0.1 s passed.
   a. Determine the acceleration of the baseball in g's. (45.6 g's)
   b. Determine the force that the pitcher applied to the ball. (51.4 N)

3.2. Imagine that you are in a helicopter, looking down on a sleigh that is being pulled by three horses (red, black, and white). You see the red horse pulling with a force of 75 lb directed 25° to the left of straight ahead. The white horse is pulling with a force of 30 lb directed at 10° to the right of straight ahead. You observe that the 200-lb sleigh is accelerating at 1.5 ft/s^2, straight ahead. You are aware that the coefficient of friction between the sleigh and the snow is 0.7. Assume that the only forces to be considered are the three horses and friction. In what direction and with what force is the black horse pulling? ($F_{\text{black horse}}$ = 58 lb at 27° to the right of straight ahead)

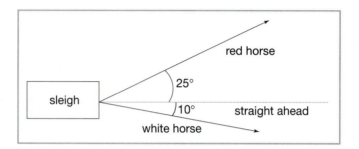

3.3. The individual in the figure is pushing a 50-lb box across the floor. The direction of the individual's force is 30° below the horizontal. The coefficient of friction between the floor and the box is 0.3. Find the magnitude and direction of the push such that the box will accelerate to the right at 1.5 ft/s^2. (24.2 lb)

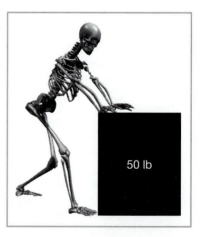

3.4. A 115-lb person on an amusement park ride is moving around a circle of radius 20 ft at 60 ft/s.
   a. Determine the net force acting on the person. (646.88 lb)
   b. Express the person's acceleration in g's. (5.6 g's—extremely dangerous)

3.5. A boy is swinging his books around his head as shown in the figure. The 15-lb books are moving in a horizontal 3.4-ft-radius circle around his body at 2 revolutions per second. He then lets go of the strap, and the books fly off.
   a. Before the books are released, how far do they travel each time they go around? (21.4 ft)
   b. Before the books are released, how fast are they moving? (42.7 ft/s)
   c. Before the books are released, what are the magnitude and direction of the force that the person must exert on them? (251.6 lb, pulling in)
   d. Express the acceleration of the books in terms of g's. (16.8 g's)

3.6. Explain why a child must run to keep up with a quickly walking adult.

3.7. A merry-go-round has a radius of 5 m and makes 5 revolutions every minute (5 rpm). A 150-lb (667 N) parent is holding onto a pole while standing next to a child riding one of the horses.
   a. Determine the force (magnitude and direction) that the parent exerts on the pole. (21 lb (93 N), radially out)
   b. How fast, in rpm, would the wheel have to turn so that the parent would experience an acceleration of 1.5 g's? (16.3 rpm)

3.8. It has been suggested that, at some time in the future, special-purpose medical facilities will be located in orbit around the earth. A person in such a facility would seem to be weightless and thus could float around, perhaps being mobile even though the skeletal muscles are not functioning. There would be little or no strain on the person's heart, as the weight of the blood being pumped up to the brain would be negligible. It is the intent of this problem to focus your attention on such a facility. (*Note*: The numerical value of g (the acceleration of gravity) depends on the altitude of the object, but at the surface of the earth (radius of the earth = 3981 miles), the accepted value of g is 9.8 m/s^2.

   a. Calculate the magnitude of g at an altitude of 250 miles above the surface of the earth. (8.7 m/s².)
   b. How fast, in miles per hour, would an object have to be moving to seem to be weightless at that altitude? (17×10^3 miles/hr.)

3.9. Explain why you feel pressed down into the seat, as if you had become heavier, on a roller coaster that is passing quickly through a low point between two steep sections.

3.10. Explain why the particulate material in a fluid-filled test tube collects at the bottom of the tube when it is centrifuged.

3.11. Roller coasters ordinarily travel on top of the tracks. Explain how a roller coaster can travel, without falling, on the bottom of a track at the top of a vertical curve. Would this trip be safer if the roller coaster moved faster or slower?

3.12. A gymnast is performing on a horizontal bar. She is executing vertical circles with her body out at arm's length.
   a. Describe the effect on her arms at the top of a swing and at the bottom of a swing when she is moving very slowly. (compression when at top, extension when at bottom)
   b. Is there a speed at which she could be going around the bar such that her arms would be neither in compression nor in tension? Explain.

3.13. Give an example (other than the patella-quadriceps tendon) of a pulley—such as a situation within the body.

3.14. Explain why it is easier, while standing, to pull a coffee table across a floor than to push it.

3.15. We all realize that a person is more likely to slip when walking up an incline than when walking on a horizontal surface. But why?

3.16. A person stands on a bathroom scale on a floor just outside an elevator. He notices that the scale reads 200 lb. He then steps into an elevator, taking the scale with him. He stands on the scale and notices that the scale reading is 210 lb as the elevator starts to move. Determine the magnitude and direction of the acceleration of the elevator at that instant. (1.6 ft/s², up)

3.17. A 180-lb physicist takes a bathroom scale along on a Ferris wheel ride. The wheel rotates at constant speed in a vertical plane. The physicist sits on the scale and notices that it reads 50 lb at the top of the circle. Determine the scale reading at the bottom of the circle. (310 lb)

MOMENTUM AND IMPULSE

There is an alternative way of expressing the second law that is helpful in explaining certain situations involving the human body. If we substitute the basic definition of acceleration,

$$a = \frac{\Delta v}{\Delta t}$$

into the second law, we get

$$F = m \frac{\Delta v}{\Delta t}$$

Multiplying both sides by Δt (the time interval during which the force is applied) yields the desired equation:

$$F \, \Delta t = m \, \Delta v$$

This equation is valid if the force is constant or if the force is replaced by its average value, $\overline{F}$. In any problem in which the force is not constant, you should use $\overline{F}$ rather than F in the equation.

The left-hand side is called the **impulse**, and the right-hand side defines the **change in momentum**:

$$\text{Impulse} \equiv F \, \Delta t$$
$$\text{Change in momentum} \equiv \Delta p = m \, \Delta v$$

The equation above is extremely important because it shows an explicit relationship between the magnitude of the average force that acts on an object and the time interval during which the force acts. In many situations, the right-hand side of the equation represents a constant value. For example, if the initial velocity and the final velocity are fixed, the right-hand side will be constant, no matter how much force is applied. There are many examples:

1. A car is moving at a given speed (e.g., 30 miles per hour) and eventually comes to a stop: $v_1 = 30$ mi/h, $v_2 = 0$. The right-hand side will have the same value whether the car coasts to a stop, the brakes are applied, or the car hits a tree.

2. A person jumps from a bridge. She will reach a certain maximum velocity as she falls and then will stop: $v_1 = v_{max}$ and $v_2 = 0$. The right-hand side will have the same value whether she is attached to a bungee cord, attached to a steel cable, or hits the ground.

3. A person is hit in the face by someone else's fist: $v_{face,1} = 0$, $v_{face,2} = v_{first}$. The right-hand side will have the same value whether the face is held rigidly up to the punch or the target falls away from the punch.

In each of these three examples, the right-hand side of the basic equation is constant. Therefore, the left-hand side must also be constant; that is, the product $F \, \Delta t$ is fixed. However, the magnitude of F depends on the magnitude of Δt. So, for example:

1. If the car coasts to a stop, Δt will be large; therefore F will be small. If the car hits a tree, Δt will be small, and F will be large.

2. If the person is attached to a bungee cord, it will stretch, and Δt will be large, resulting a small F. If she is attached to a steel cable (no stretch) or hits the ground, Δt will be very small, resulting in a large F.

3. Falling away from a punch increases the Δt and thus reduces F.

EXAMPLE 1.36

Football, tennis, baseball, soccer, golf, archery, and billiards are very different from one another, but they have at least one aspect in common. One of the goals in each of these sports is to cause an object (ball or arrow) that is initially either at rest or moving "in the wrong direction" to move as fast as possible "in the correct direction." Therefore the player wants to impart a large Δv to the object. Consider an archer. The initial velocity (v_1) of the arrow (just as the string is released) is zero. The archer wants to give the arrow as large a velocity (v_2) as possible, consistent with its going in a specific direction. Thus the archer wants to make the arrow's change in momentum as large as possible. This may be accomplished by making the force very large. Unfortunately, it is difficult to control a large force, and the arrow may travel far but not in the right direction. Reexamining the equation shows that the momentum may be changed, not only by increasing the force, but also by increasing Δt, the time interval during which the string is in contact with the arrow. This is the idea behind follow-through. The archer tries to keep the string in contact with the arrow as long as possible. This is accomplished by the archer's holding the bow in a fixed position until the arrow hits the target. Of course, once the arrow leaves the string, the bow has no effect. Holding the bow for such an extended time merely ensures that the string-arrow contact time will be as long as possible. Thus we have a large change in momentum while not requiring so large a force as to be uncontrollable. The same analysis could be used to describe follow-through by someone playing billiards, for example.

There are other situations, as in baseball and golf, in which the time interval for contact between the bat and the ball (or the club and the ball) is extremely short in comparison to the entire swing. In such cases, follow-through does not affect the change in momentum of the ball but is important to the proper biomechanical function of the body.

The following example shows how the momentum-impulse relationship can be used to understand the use of seat belts and air bags in preventing injuries associated with automobile collisions.

EXAMPLE 1.37

In the section on motion, we analyzed a 3000-lb car stopping from a speed of 30 mi/h within a distance of 1 ft. In that discussion, we showed that the force that stopped the car amounted to about 93,000 lb. Now let us consider what will happen to a 110-lb (m_p = 50 kg) person who is a passenger in the car.

Solution

Notice that both the car and the passenger are initially moving at 30 mi/h and both are eventually at rest. Thus both have the same change of velocity:

$$\Delta v_c = \Delta v_p \equiv \Delta v$$

Let us determine the change in momentum of the passenger and of the car:

$$m_p \, \Delta v = m_p(v_2 - v_1)$$
$$= (50)(0 - 13.4)$$
$$m_p \, \Delta v = -670 \text{ kg} \frac{\text{m}}{\text{s}}$$

Following the same steps, we have

$$m_c \, \Delta v = -1.82 \times 10^4 \text{ kg} \frac{\text{m}}{\text{s}}$$

The object of this example is to determine the magnitude of the external force that acts on the passenger and on the car and then to determine how any destructive effects may be lessened.

Case 1: The driver's foot is taken off the accelerator, and the car gradually slows to a stop in 25 s. Since we have already determined the change in momentum of the person, we can use the impulse-momentum relation to calculate the magnitude of the external force that acts on him.

(Note: Since we are interested only in the magnitude of the force, the minus sign will be ignored.)

$$\overline{\mathbf{F}} \, \Delta t = m \, \Delta v$$

We know that

$$\overline{\mathbf{F}}_p (25) = 670$$
$$\overline{\mathbf{F}}_p = 26.8 \text{ N}(= 6 \text{ lb})$$

This average force of 6 lb acting on the person would certainly not cause damage, in fact it would hardly be noticeable. It could easily be related to friction between the person and the seat.

Case 2: The car collides with an immovable object and comes to a stop in 0.1 s. Using the same approach, we now calculate the size of the forces that act on the car and on the passenger. Keep in mind that the changes in momentum of the passenger and of the car are the same as they were in Case 1.

$$\overline{\mathbf{F}}_c (0.1) = 1.82 \times 10^4$$
$$\overline{\mathbf{F}}_c = 1.82 \times 10^5 \text{ N} (= 4.1 \times 10^4 \text{ lb})$$

This average force acting on the car is so large that it is no wonder that the front end of the car may be totally destroyed.

To determine how much force will act on the passenger, we must know how much time it takes for him to come to a stop. He will continue moving at 30 mi/h until an external force acts to slow him down. Before the introduction of seat belts and air bags, this force would have been produced by the windshield. As the passenger hit it and was brought to a stop, the windshield deformed and was perhaps broken. Experiments have shown that, when hit by a solid object moving at 30 mi/h, a windshield will bulge outward by 15 cm. We can use these data to determine how much force acted on the passenger. We will use the momentum-impulse equation again:

$$\overline{F} \, \Delta t = m \, \Delta v$$

We already know that the right-hand side of the equation equals 670:

$$\overline{F} \, \Delta t = 670$$

To determine the force, we must know how much time it took for the windshield to stop the passenger. We will use the same approach as on page 50. We used two basic equations:

$$\vec{D} = \vec{v}\,\Delta t$$

$$\vec{v} = \frac{1}{2}(\vec{v}_1 + \vec{v}_2)$$

We have already shown that the average velocity will be 22 ft/s (= 6.7 m/s) and that the displacement of the passenger, while in contact with the windshield, will be 0.15 m. Thus Δt will be 0.022 s(= 22 ms).

The average force exerted by the windshield can be determined:

$$\bar{F} = 30 \times 10^3\,\text{N} \ (=\text{almost}\ 7 \times 10^3\,\text{lb})$$

An average force of 7000 lb applied to a person's head will certainly cause damage and may be fatal.

If a seat belt were used to stop the passenger, a very different process would occur. The seat belt is made of a material (e.g., nylon) that will stretch when a force is applied. The seat belt is not attached solidly to the car's frame; it is designed so that when a force it applied to it, the belt will slip through the attachment device. **These two characteristics, the stretching and the slipping of the belt, result in an increase in the amount of time that passes before the passenger's velocity is brought to zero. This increased time interval results in a smaller force and thus less damage experienced by the passenger.**

EXAMPLE 1.38

An air bag has several characteristics that reduce damage to the passenger:

1. The bag expands to meet the person.
2. The bag is filled with a gas that will compress when the passenger hits it.
3. The air bag presents a much larger surface than the seat belt over which the interaction with the passenger may be distributed.

Discussion

1. Since the bag expands to meet the passenger, a force will be exerted on the passenger sooner than would occur if the passenger hit the windshield. Thus Δt (the time interval over which the force is exerted) will be longer, and hence the force exerted on the passenger will be less.
2. As the bag compresses, it slows the passenger more gradually than did the seat belt or windshield (the time interval is longer). Once again, because of the momentum-impulse relationship, the passenger will experience less force.
3. Since the force is spread over a much larger area, there is less effect on any single part of the passenger's body, and less damage is done.

Experiments have shown that it takes about 0.045 s for a person to stop from a speed of 35 mi/h when an air bag is used. Using this number, we can determine that the average force exerted on the passenger will be 14.9×10^4 N (= 3.3×10^3 lb). This force, half as large as when the passenger hit the windshield, is very large, but remember that it is spread over a large area. The area of impact between the passenger and the air bag could reasonably be 2 ft² (= 288 in²). **Thus the average force per unit area exerted on the body would be about 11 lb/in². This would not cause damage to the body.**

Problem Set 4

4.1. Explain why there is less potential damage to the body when a person lands with flexed knees rather than stiff legged.

4.2. Explain how a "crumple zone" in the front end of an automobile might protect a person who is involved in a head-on collision.

4.3. Explain why it is so important to follow through in hitting a tennis ball.

4.4. We have shown that if a 110-lb passenger in a car that is moving at 30 mi/h hits the windshield, a large force (~7000 lb) will be exerted. How much force would be exerted on such a passenger in a car that is moving at 15 mi/h (6.7 m/s) who hits the windshield frame rather than the windshield itself? Experiments have shown that such a frame can distort only about 2 cm. (12.6×10^3 lb)

4.5. Two fullbacks, each weighing 250 lb, run at each other at 10 ft/s. They collide and come to a stop in 50 ms. Determine the amount of force that acted on either of them. (1.6×10^3 lb)

2

ANGULAR MOTION AND TORQUE

ANGULAR MOTION

$\mathbf{W}$e have been concentrating on the linear motion of an object, such as a box sliding on a floor, a car driving on a road, an object falling. Newton's second law, as we have stated it, relates an object's speeding up and slowing down to the external forces. We have also shown how the Second Law can be used, with the introduction of centripetal force, to deal with an object's velocity changing direction. However, a great deal of the motion of the human body involves angular (or rotational) motion, not linear motion. For example, as you walk, your arms swing in circular arcs, and your feet also move along arcs. Even lifting something may involve your hands moving along an arc around your elbow or shoulder. Rotational motion is very important in the analysis of many sports-related activities, such as the motion of a pitcher's hand or a bicyclist's feet. Bending over and straightening up are also examples of rotation.

One way to approach the analysis of rotational motion is by analogy to linear motion. The basic quantities that we have used in linear motion are displacement (e.g., Δx in meters or feet), velocity (e.g., v in meters per second or feet per second), acceleration (e.g., a in m/s^2 or ft/s^2), mass (e.g., m in kilograms or slugs) and force (e.g., F in newtons or pounds). Later, we shall define another important quantity, kinetic energy (KE), which will be measured in joules (J) or pound feet (lb ft).

Now, by analogy to these linear motion quantities, we will define the new quantities to be used in dealing with rotational motion. The quantity that plays the role of linear displacement is the **angular displacement** ($\Delta\theta$). The angular displacement is defined as the angle through which the object has rotated or the angle representing the arc through which it has moved. However, the angular displacement is not measured in degrees, as we might expect, but in radians (rad). Radians and degrees may be converted by using the following relation:

$$180° = \pi \text{ rad}$$

The quantity that plays the role of mass is the **moment of inertia** (I). It is not as simple as mass because it depends on the distribution of the mass (the shape of the object) as well as the amount of mass that the object contains. This makes the analysis of rotational motion more complicated than linear motion because, while we can reasonably expect that the mass of an object will not change while it is moving, it is certainly possible for an object's shape, and therefore its mass distribution, to change while it is moving. For example, consider a gymnast's body as he moves through the air after completing a routine on a set of parallel bars or a figure skater's arms as she spins around. In each case, the distribution of mass will change as the person extends or retracts arms and legs and tucks into a ball or straightens out. Thus while m is almost always constant, it is very easy for I to change. Notice that it is possible for an object's moment of inertia to change as a result of internal forces. Thus you could change the moment of inertia of your body using your muscles. This is very important in many sports such as skating, skydiving, and high jumping.

The basic definition of the moment of inertia of an object of mass m whose center of gravity is moving around an arc of radius r is

$$I = mr^2$$

This expression can be used unless you are given other, more specific information in a problem. You can expect that the moment of inertia of any object that you need in this course will be given to you.

EXAMPLE 2.1

As an ice skater starts to spin, she usually holds her arms out from her body. Thus her moment of inertia will be relatively large. After she starts spinning, she pulls her arms in close to her body. This decreases her moment of inertia. As a result of this, her angular velocity increases. When she wants to slow the spinning down, she extends her arms, thus increasing her moment of inertia. Her angular velocity then decreases.

Angular quantities and their analogous linear quantities are shown in the table. Kinetic energy will be discussed later in the course.

QUANTITY	LINEAR	SI UNITS	ANGULAR	SI UNITS
how far	linear displacement (x)	meters (m)	angular displacement (θ)	radians (rad)
time	T	seconds (s)	t	seconds (s)
motion	linear velocity (v)	m/s	angular velocity (ω)	rad/s
acceleration	A	m/s^2	angular acceleration (α)	rad/s^2
quantity	mass (m)	kg	moment of inertia (I)	kg m^2
force	F	N	torque (Γ)	m N
kinetic energy	KE	J	KE	J

The basic definitions will also be found by analogy:

LINEAR	ROTATION
$v = \dfrac{\Delta x}{\Delta t}$	$\omega = \dfrac{\Delta \theta}{\Delta t}$
$a = \dfrac{\Delta v}{\Delta t}$	$\alpha = \dfrac{\Delta \omega}{\Delta t}$
$F = ma$	$\Gamma = I\alpha$
$KE = \dfrac{1}{2}mv^2$	$KE = \dfrac{1}{2}I\omega^2$

The basic angular quantities $(\Delta\theta, \omega,$ and $\alpha)$ are related to the corresponding linear quantities $(\Delta S, v,$ and $a)$ by the following general expression:

$$\text{angular quantity} = \frac{\text{linear quantity}}{\text{radius of the arc}}$$

EXAMPLE 2.2

$$\Delta\theta = \frac{\Delta S}{r}$$

where ΔS is the distance, **measured along the arc,** through which either the object moved, such as a person's foot during a stride, or the object rotated, such as a person's head during a side-to-side shake.

EXAMPLE 2.3

$$\omega = \frac{v_{tangential}}{r}$$

where $v_{tangential}$ is the speed of the object or a point on the object measured around the arc, that is,

$$v_{tangential} = \frac{\Delta S}{\Delta t}$$

EXAMPLE 2.4

$$\alpha = \frac{a_{tangential}}{r}$$

where $a_{tangential}$ is the acceleration of the object, or a point on the object, measured around the arc, that is,

$$a_{tangential} = \frac{\Delta v_{tangential}}{\Delta t}$$

Notice that this is not the same as the centripetal acceleration that is radial (i.e., directed toward the center of the arc), not tangential (i.e., directed around the arc). Another important difference between centripetal acceleration and tangential acceleration is that while the former involves a change in only the direction of the velocity, the latter involves a change in the magnitude of the velocity (i.e., the object will be speeding up or slowing down as it goes around).

EXAMPLE 2.5

Calculate the angular velocity (ω) of a student sitting in a classroom in Hartford, Connecticut.

Solution

1. Definition of ω:

$$\omega = \frac{\Delta \theta}{\Delta t}$$

2. We know that the person will travel around a complete circle during 24 hours as the earth turns. Since a complete circle corresponds to 360°,

$$\Delta \theta = 2\pi \quad \Delta t = 24 \text{ hours} = 86400 \text{ s}$$

3. $\omega = \dfrac{2\pi}{86400} = 7.27 \dfrac{\text{rad}}{\text{s}}$

EXAMPLE 2.6

At the end of the windup, a pitcher's hand is stationary, but it is moving at 50 miles/hour at the instant that the ball is released. The pitcher's arm is 32 inches long. Assume that the ball moved 180° on the arc of a circle (centered at the pitcher's shoulder) as it was thrown, and calculate:

1. Its angular acceleration
2. The amount of torque (Γ) acting on it
3. Its kinetic energy (KE) as it leaves the pitcher's hand

Solution

1. We are given the following information:

$$v_1 = 0, \qquad v_2 = 50 \text{ miles/hour} = 22.352 \text{ m/s},$$
$$r_{path} = 32 \text{ inches} = 0.8128 \text{m}, \qquad \Delta\theta = 180° = \pi\text{rad}$$

2. We need angular acceleration, α. The definition of α is

$$\alpha = \frac{\Delta\omega}{\Delta t}$$

3. We need $\Delta\omega$:

$$\Delta\omega = \omega_2 - \omega_1$$

$$\omega_2 = \frac{v_2}{r} = \frac{22.352}{0.8128} = 27.55 \frac{\text{rad}}{\text{s}}$$

$$\omega_1 = 0$$

$$\Delta\omega = 27.5 \frac{\text{rad}}{\text{s}}$$

4. We need Δt:

$$\overline{\omega} = \frac{\Delta\theta}{\Delta t}$$

Assuming that the angular acceleration is constant, we have:

$$\overline{\omega} = \frac{\omega_1 + \omega_2}{2}$$

$$\overline{\omega} = \frac{0 + 27.5}{2} = 13.75$$

$$13.75 = \frac{2\pi}{\Delta t}$$

$$\Delta t = 0.457 \text{ s}$$

5.
$$\alpha = \frac{\Delta\omega}{\Delta t}$$

$$\alpha = \frac{13.75}{0.457} = 30.1 \frac{\text{rad}}{\text{s}^2}$$

6. We need the basic equation involving torque (Γ):

$$\Gamma = I\alpha$$

7. We need I, the moment of inertia. Since nothing about this was specified in the problem, we can use the basic definition:

$$I = mr^2$$

As we have pointed out earlier, the mass of a baseball is 115 g. ($= 0.115$ kg). We have already determined the radius of the arc:

$$I = (0.115 \text{ kg})(0.8128 \text{ m})^2$$
$$= 0.0760 \text{ kg m}^2$$

8.
$$\Gamma = I\alpha$$

$$\Gamma = (0.0760)(30.1)$$
$$= 2.29 \text{ m N}$$

9. We need the basic equation for angular kinetic energy (KE)

$$KE = \frac{1}{2}I\omega^2$$

$$KE = \frac{1}{2}(0.0760)(27.55)^2$$

$$= 28.8 \text{ J}$$

As can be seen from the table relating linear to angular quantities, whereas force is the cause of linear acceleration, torque is the cause of angular acceleration. Thus force and torque are analogous quantities. As we have seen before, if an object is in equilibrium, the sum of the external forces must equal zero. However, this was limited to linear motion. If we include the possibility of angular motion, for example, a person bending over, flexing the forearm or leaning sideways, we must expand the criteria for equilibrium. We must include the statement that the sum of the external torques is also equal to zero. Torque will therefore play a central role in our understanding of mechanics of the human body.

The terminology used in this field of study can lead to confusion. While the term "torque" will be commonly used in this text, the term "moment" (which means the same as torque) is also commonly used.

PROBLEM SET 5

5.1. During a tennis serve, the center of gravity of the player's arm/racket/ball (total mass $= 4.0$ kg) moves on a circular arc of radius 0.7 m. It starts from rest, reaches a maximum speed of 35 miles/hour (15.6 m/s) at $\theta = 100°$, and then comes to a stop at 240°. During Δt_1 (the first 100° of the path), one group of muscles is acting to pull the arm forward, thus increasing its speed. During Δt_2 (the final 140° of the path), a different group of muscles is acting to pull the arm backward, thus decreasing its speed. Assume that the racket's angular acceleration is α_1 during Δt_1 and is α_2 during Δt_2. The moment of inertia of the arm/racket is 1.1 kg m^2.
   a. Calculate the maximum angular velocity of the arm/racket/ball. (22.4 rad/s)
   b. How much time does it take for the arm/racket/ball to reach this maximum velocity? (0.16 s)

     d. How much torque was applied by the first group of muscles? (158 m N)

     e. How much torque was applied by the second group of muscles? (113 m N)

5.2. Explain why a high diver tucks into a ball while falling and then straightens out just before hitting the water.

5.3 Explain how a figure skater controls the speed at which he or she spins.

TORQUE : INTRODUCTION

In the preceding section, we used torque to analyze rotational motion problems. We did not really describe torque; we only used it as represented by a formula. To deal with situations relating to the body, in particular those related to posture, muscle strain, tendon damage, and many other types, a more detailed knowledge of torque is required.

Suppose that you wanted to move a ladder that was on the ground so that it would be leaning up against a wall. You could walk to the end closer to the wall, place your foot at the end so that the ladder could not slide (the pivot point), and then bend over and pull on the ladder. If you were strong enough, you could rotate the ladder so that it would lean against the wall. It doesn't take much experience to learn that it is a lot easier (i.e., less force is required) to walk to the end farther from the wall, lift that end up, and then walk toward the wall, lifting the ladder as you go. Eventually, it will be leaning up against the wall. Why is the second approach so much easier than the first? It must have something to do with where the force is applied. If the force is applied close to the pivot point, a great deal of force is needed to lift the ladder. If the force is applied far from the pivot point, less force is needed. It is this link between the location (and direction) of a force and its magnitude that is central to the concept of torque.

The **torque** (Γ) associated with a force is the product of the magnitude of the force (F) multiplied by the lever arm (LA) associated with the force:

$$\Gamma = (F)(LA)$$

The units associated with torque will therefore be the following:

SI: newton-meter (N-m)
USA: pound-foot (lb-ft)

Just as the terms "moment" and "torque" are used interchangeably, so too are "moment arm" and "lever arm. "

The **lever arm** is the distance measured from the **axis of rotation**, perpendicular to the **line of action** of the force. The **line of action** of a force is an infinite line drawn on the FBD that coincides with the arrow representing the force. The **axis of rotation** is a line about which the object is rotating or might rotate.

QUESTION:
Describe the axis of rotation for each of the figures.
Hammer:

Wrench:

Tire iron:

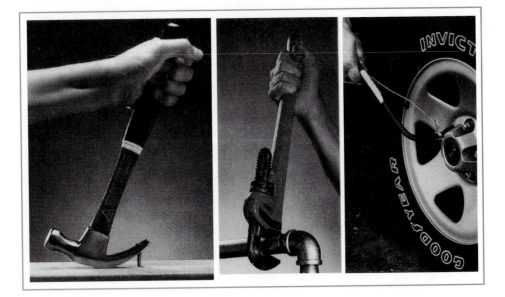

QUESTION:
Why is the person holding each of the tools near its end?

QUESTION:
Could the plumber increase the torque applied to the wrench by his hand without increasing his force?

The idea of an axis of rotation may also be applied to the body. If I raise my forearm, it will rotate about a line passing through my elbow. If I bend over, my trunk rotates about a line passing through my hips, more precisely the joint between the fifth lumbar vertebra and the sacrum (L5-S1).

Torque is a vector quantity, and so a means of specifying direction must be agreed on. One convention is to specify direction of a torque by determining whether that torque will produce a clockwise (CW) or counterclockwise (CCW) rotation. In the examples in this book, I will always associate CW torques with a positive sign and CCW torques with a negative sign. **This particular sign convention is not necessary, but you must specify which torque (CW or CCW) will be positive in any analysis that you carry out.**

From the formula for torque, we can see that for a given amount of torque, for example, 50 ft lb, the amount of required force must increase as the length of the lever arm decreases.

TORQUE	LEVER ARM	FORCE
50	10	5
50	5	10
50	1	50
50	0.5	100
50	0.1	500

These numbers illustrate an idea mentioned earlier:

A force that is applied close to an axis of rotation will produce a smaller torque than if it were applied farther from the axis of rotation.

In the figure above, a very small tugboat is pushing on the bow of the large ship to turn it. To make effective use of its limited force (push), the tugboat is positioned as far as possible from the stern of the ship. Notice also that the tugboat is aimed so that its force is perpendicular to the axis of the ship. Both of these factors contribute to making the lever arm as long as possible. This produces the largest possible torque, given the limited amount of force that may be exerted by the tugboat.

It is generally true that the body's internal forces, those associated with muscles and tendons, are applied close to the axes of rotation. For example, the biceps muscle inserts onto the radius (one of the two bones that make up the forearm) just below the elbow. The deltoid muscle inserts onto the humerus (the bone that is the upper arm) just below the shoulder. The quadriceps tendon inserts onto the tibia (one of the two bones that make up the lower leg) just below the knee. Since these forces are applied close to an axis of rotation, they are associated with short lever arms. For any of them to produce a large torque, that muscle must represent a large force. So, as we shall see later, the tension in the deltoid muscle will be much larger than the weight of an outstretched arm that it is supporting. We have already seen (in our discussion of components) that a large force in a muscle will cause a large amount of force compressing the joint, often leading to pain and breakdown of the synovial tissue and/or the bones themselves.

There is a tradeoff that is beneficial to the body: strength for speed. A small amount of contraction of the deltoid muscle can cause a much larger amount of motion as the arm is forced to swing vertically. This multiplication in motion is directly related to the fact that the deltoid muscle inserts close to the axis of rotation, the shoulder. Thus the same structure that necessitates large forces in the muscles provides for the large amount of resulting motion of our limbs.

EXAMPLE 2.7

The knee is made up of three bones: the femur (upper leg), the tibia (main bone in lower leg), and the patella (kneecap). The quadriceps muscle, the large muscle on the front of the thigh, is the major extensor of the knee. When it contracts, it pulls on the quadriceps (or transpatellar) tendon that passes over the patella and then is connected to (inserts onto) the tibia. Thus when the quadriceps muscle contracts, it pulls on the tibia so as to make it rotate toward full extension (straight rather than bent). Thus a torque is present.

The force is the amount of tension in the tendon, and the lever arm is the perpendicular distance from the tendon to the rotation axis (where the femur and tibia "come into contact"). The lever arm is not a fixed distance because the patella slides along the femur as the tibia rotates. The motion of the patella causes the lever arm to increase (as the patella slides up) or decrease (as the patella slides down). There is a groove (patellar surface) in the end of the femur in which the patella slides. This groove is relatively deep at the end of the femur and becomes shallow toward its upper end. Since the patella slides within this groove as you flex and/or extend the knee, the patella-joint distance (lever arm for the tendon) must increase as the patella slides upward toward the top of the femur. Thus near full extension of the leg, the tension in the quadriceps tendon is less than it is near full flexion because its lever arm has become longer. This may be seen from the following three figures, which are X-rays of the author's knee as it is flexed from 0° to 45° and then to 90°. The patella is displaced down along the femur, and correspondingly, the lever arm decreases from one figure to the next (see p. 93).

As was pointed out above, the length of the LA for the quadriceps tendon changes as the lower leg flexes. A similar situation occurs at the shoulder (Norkin and Levangie, 1992, p. 43). As the upper arm (humerus) moves from a vertical to a horizontal position (exhibits eccentric motion), the LA associated with the deltoid muscle changes. The shape of the upper end of the humerus is such that the muscle's LA is greater when the arm is 60° from the vertical than at any other point.

EXAMPLE 2.8

Skeletal muscle reacts to applied stress by building more muscle mass. This characteristic is exploited when one uses exercise equipment to increase muscle tone or to bulk up muscles. A difficulty is encountered, however, in the choice of equipment to be used. The choices are free weights (inexpensive, easily available), spring-based equipment (moderately expensive, easily available), and cam-based equipment (very expensive, available only in workout centers, gyms, etc.). Each of these types of equipment has advantages and disadvantages. However, we are

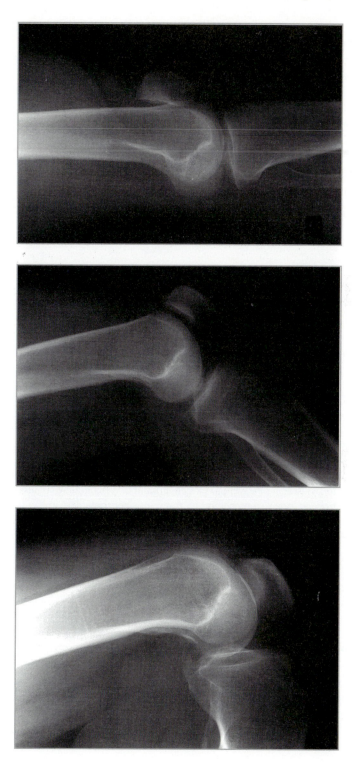

going to be concerned only with how well the force exerted by such a piece of equipment correlates with the maximum force that may safely be exerted by the muscle.

One of the characteristics of skeletal muscles is that the maximum strength varies through the range of motion of the muscle. Thus such a muscle may be weak at full extension or at full contraction and strong somewhere in between (usually near the midpoint). It is also known that muscles build up, or add muscle mass, as a result of being strained. This presents a problem for someone who is attempting to build up such a muscle. If the muscle is forced to work against a force that is sufficiently large to maximally strain it at its strongest point, the muscle may be damaged near full extension or full contraction. If the muscle is forced to work against a force that safely strains it near the endpoints, the muscle will not be sufficiently strained at it strongest point, near the midpoint of the range.

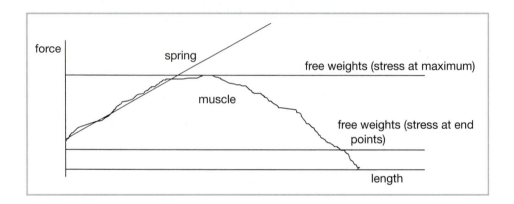

Free weights present a major difficulty. The amount of force is constant over the range of motion of the muscle. If one were to pick an amount of weight that will maximally strain the muscle at its strongest point (represented by the upper horizontal line in the figure), the midpoint, then the muscle may be damaged near its maximum contraction or extension, where it is much weaker. If one were to pick a weight that would not damage the muscle near the endpoints of the range of motion (represented by the lower horizontal line in the figure), then the muscle may not be maximally strained at the midpoint, where it is at its strongest.

Spring-based equipment presents a varying force. The force gets larger as the spring becomes more stretched (represented by the diagonal line in the figure). Thus the force applied to the muscle increases as the muscle lengthens from maximum contraction to the midpoint of the range. This matches the strength curve of the muscle during the first part of the range of motion. However, as the midpoint is passed, the muscle becomes weaker, yet the force applied by the spring continues to increase. Thus damage to the muscle may result.

Cam-based machines seem to address this problem. The photograph on page 95 shows a Cybex leg curl machine. The cam is the oblong disk in the center of the figure. To use the machine, the person lies prone on the mat, head down and to the left. As she contracts her hamstring muscles, her ankles push up on the small pad near the chair, causing the cam to rotate CCW. The cable is thus pulled down, raising the weights.

At any time during the exercise, the cam is not accelerating, and therefore the sum of the torques acting on it must be zero. There are only two torques playing a role (assuming that the axle is the axis of rotation). The torque associated with the cable must be balanced by the torque associated with the force applied by her legs:

$$\vec{\Gamma}_{cable} + \vec{\Gamma}_{legs} = 0$$
$$|\vec{\Gamma}_{cable}| = |\vec{\Gamma}_{legs}|$$
$$F_{cable} LA_{cable} = F_{legs} LA_{legs}$$

$$F_{legs} = \left(\frac{F_{cable}}{LA_{legs}}\right) LA_{cable}$$

The design of the apparatus is such that neither the force in the cable nor the lever arm associated with the person's legs changes during the range of motion. The quantity in the parentheses above is then constant. Thus the force associated with her legs should vary in the same way as the cable's lever arm. The shape of the cam is such that the cable's lever arm is relatively small near either extreme of the range of motion and large near the midrange. This is shown by the curve with plus signs in the figure on page 96. The other curve (the squares) shows how her applied force varies.

The second curve shows that the design concept was achieved. **The force applied by the person varies through the range of motion in accordance with the capability of the specific muscles: a relatively small force at either extreme and maximum force near the midpoint.**

Center of Gravity

In trying to analyze a realistic problem, we quickly find that the number of forces is so large that either the problem cannot be analyzed or the analysis would be too difficult to

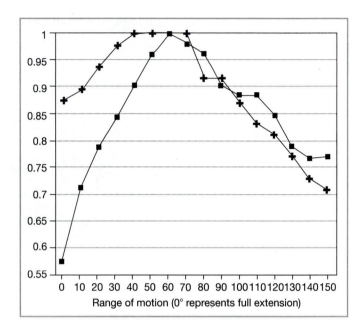

carry out. For example, consider a person who is engaged in gymnastics with a trampoline. The forces would include the following:

The force of the trampoline pushing up on the person's feet

The air resistance pushing on each part of the person's body that is in contact with the air

The force of gravity pulling straight down on each part of the person's body

All of the pushes and pulls exerted by parts of the body on each other

Fortunately, the problem may be significantly simplified. First of all, we might agree to deal with the body as a whole and not consider each part of the body separately. Then we can represent the effect of gravity on the entire body by drawing a single vertical arrow rather than a collection of many vertical arrows. **However, this arrow must be drawn through the center of gravity (cg).** To do this, we must be able to locate the cg for any given object.

For a regularly shaped object, such as a rectangular solid or a sphere, the cg is at the geometric center. For an irregularly shaped object, such as the human body, the location of the cg must be determined experimentally. For example, if we wanted to locate the cg of the 48 contiguous states, we might use the approach sketched in the next figure. Hang a map of the United States from a nail that pierces the northwest corner (near Seattle, Washington). The map must be free to rotate around the nail. After the map has stopped moving, hang a weighted string from the nail, and draw the vertical line. Repeat this, locating the nail near the northeast corner (the tip of Maine) and then perhaps somewhere near the middle of the border with Canada (perhaps on the North Dakota boundary). The three lines should intersect at a point (some place in Nebraska). That point represents the bal-

ance point or cg of the map. (*Note*: Only two lines were needed; the third line serves as a check.)

To summarize, there are three advantages to using the center of gravity:

1. The motion of the cg of an object is not influenced by any forces internal to the object. The motion of the cg is influenced by only those forces that are external to the object. So, considering the person using the trampoline, the motion of the person's cg will depend only on the forces represented by the push of the trampoline, the push of the air resistance, and the pull of the force of gravity on each part of the person's body. The various pushes and pulls within the person's body will affect the motion of a hand or of a foot but will not affect the motion of the person's cg.

2. Another result of using this model is that rather than having to deal with the force of gravity as exerted on each part of the object, we need only consider the total force of gravity on the whole object as acting on this one point, the cg. This means that we need draw only one arrow for the force of gravity, or the weight, of an object. That arrow will always point vertically down and will pass through the cg of the object.

3. Another aspect of this model (the cg of an object) is that we can deal with balance.
   a. If we were to cut a map of the United States (as shown in the figure) out of a piece of plywood, that map could be balanced on a point that is located where the strings intersect.
   b. For a freestanding person to be stable, it is necessary that the cg be located on any vertical line that passes between the person's feet.

The location of the cg depends on the geometry of an object as well as on its mass distribution. This means that I can change the location of my cg by changing my body's geometry. If I feel that I am falling over to the left, I can move my cg to the right by sticking out my right arm. If I am carrying a heavy load in front of me, I may fall over forward. I automatically compensate for this by leaning back to move my cg backward so that it is over my feet. Just as you can change the location of your cg by changing the shape of your body, you can also change another important biomechanical quantity, the moment of inertia of your body, in the same way. This was pointed out in the discussion of angular motion.

Another simplification deals with the force of air resistance. From experimental evidence, we know that the force of air resistance depends on the velocity of the object, as well as on its shape. If we were to try to analyze the motion of a person riding a bicycle, we would find that air resistance would be a major force acting on the person and could not be ignored. However, the analysis of problems that involve air resistance presents severe mathematical difficulties. Therefore we will not consider such problems in this course.

Shoulder

EXAMPLE 2.9

Consider a 120-lb person who is holding a 115-g baseball out to the side at arm's length. There will be several forces acting on the arm:

1. the deltoid muscle, acting at its insertion point, pulling up and to the left
2. the weight of the arm, acting at the cg of the arm and pointing down
3. the weight of the baseball, acting at the hand and pointing down
4. the force exerted on the arm at the shoulder, acting at the shoulder, direction unknown

Since several forces are acting on the arm, it is reasonable that there will be several torques acting on it. In previous work, we had to add up the external forces that were acting on an object. Here, we have to add up the external torques. Remember that some of them will be positive and some of them will be negative. If the object is in equilibrium (i.e., not accelerating), then not only will the sum of the external forces be equal to zero, but the sum of the external torques will also be equal to zero. These ideas will provide the basic equations for the analysis.

Procedure for analyzing equilibrium problems

1. Identify the object of interest. (*Note*: This must be an object, for example, a particular bone such as the tibia, or a part of the body that is made up of particular bones, for example, the entire right leg.)
2. List all of the forces that are acting on the object. Identify them by appropriate symbols and indicate magnitude and direction if known.
3. Make a FBD.
   a. Represent the object by an elongated line or shape.
   b. Represent each of the forces acting on the object as an arrow pointing outward from the object. It is important that each arrow is placed and oriented on the diagram so that it correctly represents the external force being dealt with. Label the forces with appropriate symbols or numerical values. Superimpose a coordinate system. Any unknown force should be repre-

sented symbolically, for example, F_{1X} and F_{1Y}. As we discussed earlier, the directions of these components may be chosen arbitrarily. If you guess wrong, the magnitude of that component will come out to be a negative number.

  c. Locate the axis of rotation (to be discussed on page 100).

4. Write the basic equations that will be used to represent the situation. These will usually come from Newton's second law for linear and for rotational motion:

$$\sum_{\text{all forces}} \vec{F}_{\text{external}} = 0$$

$$(\rightarrow +)\sum_{\text{all external forces}} F_X = 0$$

$$(\uparrow +)\sum_{\text{all external forces}} F_Y = 0$$

$$\curvearrowleft \sum_{\text{all external forces}} \vec{\Gamma}_{\text{axis of rotaion}} = 0$$

$$\curvearrowleft \sum_{\text{all external forces}} F \cdot LA_{\text{axis of rotation}} = 0$$

5. Substitute from the FBD into the basic scalar equations.
6. Solve the equations for the desired quantities.

EXAMPLE 2.10

Let us apply the procedure outlined above to the problem of the 120-lb man holding a 115-g baseball out to his side. Since the forces that we have been discussing act on his left arm, we shall pick it to be the object of interest. The figure shows his arm. There are arrows representing the deltoid muscle, the weight of the arm (acting at the cg of the arm), and the weight of the ball.

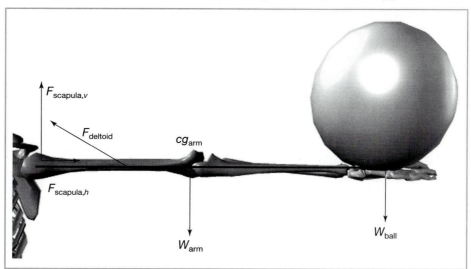

As we pointed out earlier, there will also be a contact force (F_c) exerted on the upper arm (humerus) by the shoulder (scapula). Since its direction is unknown, it will be represented on the FBD by two arbitrarily chosen components: F_h and F_v.

In making a FBD, it is necessary to take into account the location of each force as well as its direction. To do this, the object is represented by an extended line or shape rather than by a dot as before. In this example, the object of interest will be the person's arm.

The FBD is shown in the figure.

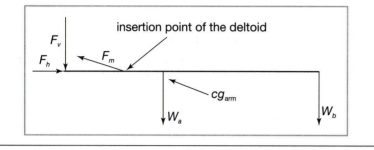

In dealing with torque, it is always necessary to refer to a specific axis of rotation. For this example, an obvious axis of rotation will be where the humerus meets the scapula (i.e., at the left end of the FBD). If you imagine that the line that represents the arm is free to rotate around that end, you can see that the two forces W_a and W_b would tend to make the line rotate CW, while the force F_m would make the line rotate CCW.

Notice that the lines of action of each of the forces F_h and F_v pass right through the axis of rotation. Each of their lever arms will therefore be zero, and they would produce zero torque. This illustrates an important point:

The torque associated with any force whose line of action passes through the axis of rotation will be zero.

From a theoretical perspective, it is possible to locate the axis of rotation anywhere. However, there are usually two locations that are particularly attractive. One of these is the "real" axis of rotation, for example, the knee, ankle, hip, or some other joint in the body. The other choice for the axis of rotation is one that results in eliminating one or more unknown forces from the torque calculation. Although the first of these two choices is easier, conceptually, I recommend the second approach, since it will result in less involved calculations. For Example 2.10, the two considerations result in the same location for the axis of rotation: the shoulder.

Consider the force W_a. Since it is a weight, its line of action would be a vertical line, as shown in the figure. Its lever arm (that must be perpendicular to the line of action) would be horizontal. It is also shown in the figure on page 101.

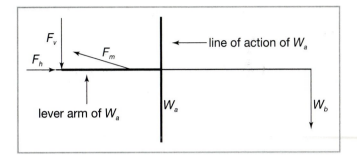

Consider the force F_m. Its line of action and its lever arm are shown in the next figure.

To proceed with the analysis, we need some quantitative information. Referring to Appendix 7, we see that we can make use of the following data:

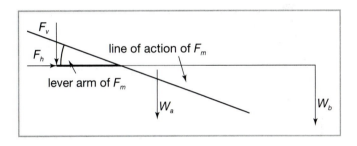

Shoulder joint to deltoid insertion: 13 cm

Angle between deltoid and axis of arm: 15°

Length of arm: 52 cm

Location of cg of arm: 46% of length of arm = 24 cm

Weight of arm: 4.8% of body weight that (assuming 120-lb person) = 5.76 lb

The mass of the ball was given as 115 g. Its weight will be 1.13 N.

We need to pick a system of units; we'll use the SI system. This will require changing centimeters to meters and pounds to newtons. Now we can label the FBD.

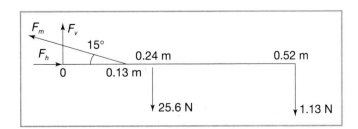

Notice that the directions of the two components of the unknown force (F_s) have been assumed to be up (F_v) and to the right (F_h), as suggested earlier.

We will now use the basic equations:

$$\sum_{\text{all forces}} \vec{F}_{\text{external}} = 0$$

$$(\rightarrow +)\sum_{\text{all external forces}} F_X = 0 \qquad \text{(the } X \text{ equation)}$$

$$(\uparrow +)\sum_{\text{all external forces}} F_Y = 0 \qquad \text{(the } Y \text{ equation)}$$

$$\curvearrowright \sum_{\text{all external forces}} \vec{\Gamma}_{\text{axis of rotation}} = 0$$

$$\curvearrowright \sum_{\text{all external forces}} F \cdot LA_{\text{axis of rotation}} = 0 \qquad \text{(the torque equation)}$$

Substituting from the FBD into the X equation, we get

$$F_h - F_m \cos 15° = 0$$

Substituting from the FBD into the Y equation, we get

$$F_v + F_m \sin 15° - 25.6 - 1.13 = 0$$

Since we now have two equations that involve three unknowns, we need a third equation: the torque equation. To write a torque equation, we must specify the location of the axis of rotation and the positive direction of rotation. It should be clear from the FBD that if the shoulder (the left end) of the arm is picked to be the axis of rotation, two forces $(F_h$ and $F_v)$ will have lever arms equal to zero. Thus both of these torques will be zero. That will certainly simplify the math, so we will pick the left end of the FBD to be the axis of rotation. Either CW or CCW could be chosen to be the positive direction. I will pick CW.

$$\curvearrowright \Gamma_h + \Gamma_v + \Gamma_m + \Gamma_a + \Gamma_b = 0$$
$$(F_h)(LA_h) + (F_v)(LA_v) - (F_m)(LA_m) + (W_a)(LA_a) + (W_b)(LA_b) = 0$$
$$0 + 0 - (F_m)(0.13 \sin 15°) + (25.6)(0.24) + (1.13)(0.52) = 0$$

This yields the following value for the force exerted by the deltoid muscle:

$$F_m = 200 \text{ N} (= 45 \text{ lb})$$

Substituting this value into the X and Y equations, we get

$$F_h = 193 \text{ N}$$
$$F_v = -25 \text{ N}$$

Notice that the Y component of the unknown force came out to be negative. This means that the original guess was wrong. This component points down, not up.

We now draw these components on a set of coordinate axes.

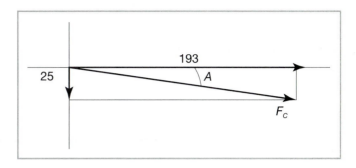

We now pick one of the right triangles and use the Pythagorean theorem and inverse tangent relation as we did earlier in the course.

$$F_{contact} = 195 \text{ N } (= 44 \text{ lb}) \text{ directed at } 7.4° \text{ below the } +X \text{ axis.}$$

Notice that the force exerted by the deltoid muscle (45 lb) is much greater than the combined weights of the arm (5.76 lb) and of the ball (0.25 lb). The relatively large amount of tension in the muscle is caused by its short lever arm. This is a common occurrence within the body. Also notice the relatively large amount of compression (F_c = 193 N = 43 lb) at the shoulder. This is caused by the force of the deltoid. A great deal of compression of a joint, as caused by the large amount of tension in skeletal muscle, is also common in the body.

Knee

Example 2.11

Consider a situation in which a 200-lb (889.6-N) person is squatting down on both legs while supporting a 300-lb (1334.5-N) weight. The person's heels will be off the ground, the lower legs will be bent forward, and the upper legs will be bent backward.

We are going to concentrate on the forces acting on one of his lower legs. The arrows in the sketch represent forces acting on his left leg.

For this particular person, we are given the following information:

1. The quadriceps tendon inserts onto the tibia at a point 4 cm down from the knee joint as measured along the surface of the tibia. The tendon makes an angle of 5° with the surface of the tibia.
2. Length of tibia: 36 cm
3. Distance from the ball of the foot to the talus-tibia contact: 12 cm

Determine the following:

1. the tension (Q) in the quadriceps tendon
2. the magnitude and direction of the force (R) that acts on the upper end of the tibia

Solution

1. The object of interest is the lower leg (including the foot).
2. The forces acting on this object are as follows:
    a. Weight$_{object}$

$$W = 5.9\% \text{ of body weight} = (0.059)(889.6 \text{ N})$$
$$= 52.5 \text{ N (acting at the cg of the lower leg)}$$

   b. Force$_{floor}$

$$F_n = 0.5 \text{ [weight of person (889.6 N)} + \text{weight of the load (1334.5 N)]}$$
$$= 1112.1 \text{ N}$$

(*NOTE*: the factor 0.5 is due to the fact that the person is standing on two feet and we are considering only the forces that act on one of them.)
   c. Tension in the quadriceps tendon: Q

d. Force at the femur-tibia contact point: R. Neither the magnitude nor the direction of this force is known. It will be represented on the FBD by two components.

3. The FBD:

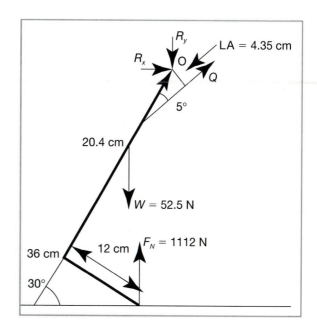

4. Basic equations:

$$\rightarrow + \ \sum F_X = 0$$

$$R_X + Q \cos 25° = 0$$

$$\uparrow + \ \sum F_Y = 0$$

$$-52.5 + 1112 + Q \sin 25° - R_Y = 0$$

I have chosen to place the axis of rotation at the upper end of the tibia (0). This choice will produce lever arms of zero length for both R_X and R_Y. Usually, the lever arm for the quadriceps tendon would have to be calculated. However, in this problem, that calculation would involve more trigonometry than is appropriate for the course. The lever arm is therefore given as 4.35 cm.

$$\curvearrowright \sum \Gamma_0 = 0$$

$$(1112)[(0.36 \cos 30) - (0.12 \cos 60)] - (52.5)(0.204 \cos 30°) - (Q)(0.0435) = 0$$

5. Calculate the components of the unknown force R:

   $Q = 6223$ N
   $R_X = -5639$ N   (the negative sign means that we guessed wrong
                            and that R_X really points to the left)
   $R_Y = 3689$ N   (the positive sign means that we guessed correctly and that R_Y
                            really does point down)

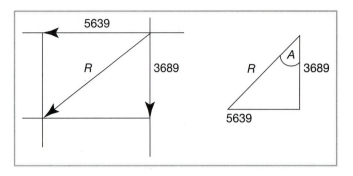

6. Determine the magnitude and direction of R:

$$R = 6.74 \times 10^3 \text{ N} (1.51 \times 10^3 \text{ lb})$$
$$A = 56.8°$$

Notice that the femur seems to be pushing to the left on the tibia. This doesn't seem possible. It is really the ligaments, the ACL in particular, that stabilize the knee (rather than the femur itself) and that are applying the force on the tibia. When a competitive weight lifter does a lift as described in this example, great stress is placed on these ligaments. This explains why weight lifters commonly wrap their knees in Ace bandages and why ligament failure (knee dislocation) is so common among weight lifters.

Also notice that the short lever arm associated with the tendon results in a large tension within the tendon that, in turn, results in a large force of compression on the joint. This lever arm would be much shorter if it were not for the patella. The patella rides on the end of the femur and the tendon passes over it. Thus the patella serves to move the tendon out, farther away from the joint. The patella serves, both by its placement and by its motion (review the discussion of knee in the section entitled "Torque: Introduction," to reduce the amount of force that must be exerted by the quadriceps tendon.

Lower Back

EXAMPLE 2.12

Consider a 150-lb person who is bending forward from the waist while holding a 10-lb weight. The angle between his back and the horizontal is 60°. The angle between his arms and the horizontal is 20°. It is known that the length of his arms is 52 cm and the length of his trunk (neck to hips) is 40 cm. Find the magnitude and direction of the force exerted by the sacrum on the lumbar vertebrae.

Solution

1. The only forces that appear in the basic equations are external forces. Since we want to determine the mag-

nitude and direction of the force that the sacrum is exerting on the lumbar vertebrae, that force must be an external force. Therefore the object of interest must include either the lumbar vertebrae or the sacrum but not both. Since the weight that is being supported is acting on the upper body (above the L5-S1 joint) and since we have information about the arms and trunk (parts of the upper body), we will try an analysis based on the object of interest being the upper body. The object of interest will be assumed to be made up of the trunk of the body and the head and arms.

2. The forces acting on this object are as follows:

W_t: the weight of the trunk, head, and neck of the body (56.5% of the body weight) = 84.8 lb = 377 N. The cg of trunk, head, and neck is 0.604 of the length of the trunk (0.242 m) down from the shoulder.

W_a: the weight of the two arms and hands (9.6% of the body weight) =14.4 lb = 64 N. The cg of arm and hand is 0.46 of the length of the arm (0.239 m) from the shoulder.

W_l: the weight of the load =10 lb = 44.5 N.

F_e: the pull of the erector spinae muscle. The insertion point of this muscle is two thirds of the way up from the hips (0.267 m up from the hips). The muscle makes an angle of 12° with the spinal column.

F_s: the force exerted by the sacrum on the lumbar vertebrae (unknown).

The data above use information from the table of weights and centers of gravity of body segments from Appendix 9.

3. The FBD.

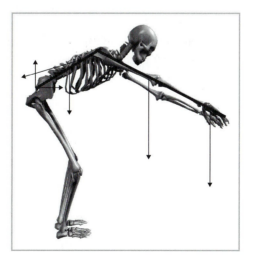

Since F_s, the force exerted by the sacrum on the lower end of the spinal column, is unknown, it is represented in the FBD by its components, F_{sh} and F_{sv} (directions assumed).

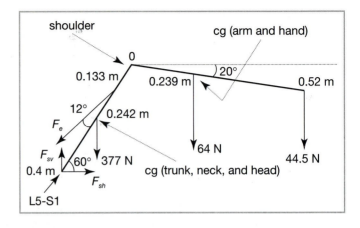

4. Notice that there are three unknowns in this problem: F_{sh}, F_{sv}, and F_e. F_{sh} and F_{sv} are the components of the force that is to be determined, and F_e is the force of the erector spinae muscle. We will consider the X and Y equations first:

$$\rightarrow + \ \sum F_X = 0$$

$$F_{sh} - F_e \sin 42 = 0$$

$$\uparrow + \ \sum F_Y = 0$$

$$F_{sv} - F_e \cos 42 - 377 - 64 - 44.5 = 0$$

$$F_{sv} - F_e \cos 42 - 485.5 = 0$$

We now have two equations, but they contain all three unknowns. So we need another equation:

$$\curvearrowleft\oplus\sum \Gamma_{\text{sacrum}} = 0$$

To ensure that none of the torques is omitted, make sure that you have the same number of torques as the number of forces shown on the FBD (six):

$$\curvearrowleft\oplus \Gamma_L + \Gamma_A + \Gamma_T + \Gamma_{F_e} + \Gamma_{sh} + \Gamma_{sv} = 0$$

Consider F_{sh}. Since its line of action passes through the axis of rotation (the sacrum), its lever arm will be zero, and hence its torque will be zero. So too with F_{sv}. Consider W_T. This force will produce a CW torque about the sacrum, and so its torque will be positive. Its line of action will be a vertical line (as will the line of action for any force that is a weight). Therefore its lever arm will be a horizontal line drawn from the axis of rotation to the line of action. The figure on page 109 shows several forces (heavy dark lines) and their lever arms (dotted lines).

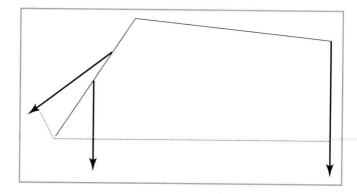

The lever arm (*LA*) for the 377-N force will be 0.158 cos 60° m. Consider the erector spinae muscle (F_e). Its lever arm will be 0.267 sin 12° m. Consider the 44.5-N force. Its lever arm will be 0.4 cos 60 + 0.52 cos 20 m.

Eventually, all of the forces will have been considered, and then the following torque equation may be written:

$$(377)(0.158 \cos 60) - (F_e)(0.267 \sin 12) + (64)(0.4 \cos 60 + 0.239 \cos 20)$$
$$+ (44.5)(0.4 \cos 60 + 0.52 \cos 20) = 0$$

$$29.783 - F_e(0.056) + 27.174 + 30.644 = 0$$
$$F_e = 1564.3 \text{ N}$$

5. We can now substitute this value into the equations for F_{sh} and for F_{sv}:

$$F_{sh} = 1046.6 \text{ N}$$
$$F_{sv} = 1647.8 \text{ N}$$

Notice that both F_{sh} and F_{sv} came out as positive numbers. That means that we guessed correctly when we drew them on the FBD.

6. We can now draw the two components of F_s on a set of Cartesian axes.

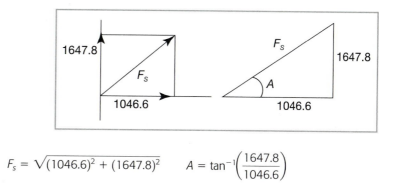

7. $$F_s = \sqrt{(1046.6)^2 + (1647.8)^2} \qquad A = \tan^{-1}\left(\frac{1647.8}{1046.6}\right)$$

Thus

$$F_{\text{sacrum}} = 1.95 \times 10^3 \text{ N} = 439 \text{ lb} \qquad A = 58° \text{ above horizontal}$$

Notice that even though the person is holding a weight that represents less than 10% of his body weight, the force compressing the lumbar vertebrae is almost 3 times larger than his total body weight.

If the 150-lb person were to stand erect with the 10-lb weight, how much force would be compressing the lumbar vertebrae? The answer is the weight of that part of his body above L5-S1(66.1% body weight = 99.15 lb) plus the weight of the load (10 lb) that is 109 lb. So if he were to bend over, pick up the weight, and then stand upright, the force compressing the lumbar vertebrae would vary from 439 lb to 109 lb. You can imagine that if this were to be done repetitively, the soft tissue between the vertebrae, and perhaps the vertebrae themselves, would eventually break down.

The lever arm associated with the erector spinae muscle is very small in comparison to the other lever arms in the problem. This results in the erector spinae muscle exerting a very large force, 1.56×10^3 N ($= 350$ lb, twice the body weight), which will have the effect of compressing the vertebrae (as well as keeping the body erect). This is the reason that employing an incorrect lifting technique, such as that described in this example, may lead to damage to one's back.

It is a curious situation that if you lift a weight by bending over while keeping your knees straight, you protect your knees but place your lower back at risk. If you keep your back straight but flex your knees, you protect your lower back but put your knees at risk.

Hip

EXAMPLE 2.13

A person who has an injured hip will usually shift his or her body toward the side that is injured. This results in walking with a limp. The goal of this example is to understand why this shift occurs.

The picture shows a 140-lb (622.8-N) person who is walking normally at an instant when only the right foot is on the ground. The right leg, which is bearing the person's weight, will automatically tilt outward at the top. It may reasonably be assumed that under these conditions, the hip abductor muscles pull at an angle of 20° from the vertical.

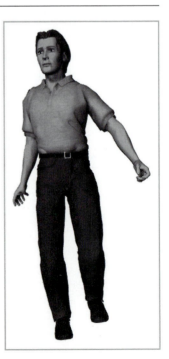

The specific data for this problem are as follows:

The length of the leg is 85 cm.

The distance from the greater trochanter to the head of the femur is 6 cm.

The horizontal distance between the greater trochanter and the cg of the body is 15 cm.

The weight of the entire body is 622.8 N.

1. Explain why the right leg automatically tilts outward.
2. Determine the magnitude of the contact (or reaction) force acting between the acetabulum and the femur.

Solution

1. Since we are to calculate the amount of force that the acetabulum exerts on the head of the femur, the femur must be part of the object of interest. If we were to pick the femur as the object of interest, we would have to deal with the weight of the femur, the forces exerted on the femur by the tibia and by the quadriceps muscle, and the forces exerted by the acetabulum and the hip abductor muscles. This would represent too many unknowns. If we include the lower leg (tibia) and the foot as part of the object of interest, then some of these forces (e.g., the tibia and quadriceps) become internal rather than external and, as such, play no role in the analysis. This simplifies the problem. So let us consider the entire right leg as the object of interest.

2. The forces that act on the right leg are as follows:
   a. Hip abductor muscles: F_m
   b. Acetabulum: F_a
   c. Gravity: W_{leg} = 15.6% of the body weight = 97.16 N (acting at the cg of the leg)
   d. The force exerted by the floor: F_f = body weight = 622.8 N

3. The FBD of forces acting on a right leg when a person is standing only on that leg follows. Notice that since we do not know the direction of the force exerted on the head of the femur by the acetabulum, we represent it as a pair of components. I have chosen to draw these components down and to the left (as shown on the FBD on p. 112).

   Since the body's cg is directly over the right foot, the angle A may be calculated by using $\sin^{-1}\left(\dfrac{0.15}{0.85}\right)$ $A = 10.2°$

4. Notice that there are three unknowns: F_m, F_{aX}, and F_{aY}. Therefore we will need three equations:

$$\rightarrow + \ \sum F_X = 0$$

$$F_m \sin 20° - F_{aX} = 0$$

$$\uparrow + \ \sum F_Y = 0$$

$$622.8 - 97.16 - F_{aY} + F_m \cos 20° = 0$$

We will use the torque equation for our third equation. To do this, we must select an axis of rotation. I will pick the greater trochanter (indicated as 0 on the FBD) to be the axis of rotation:

$$\curvearrowleft \sum \Gamma_0 = 0$$

$$(F_{aY})(0.06 \sin 44.8°) - (F_{aX})(0.06 \cos 44.8°) + (97.16)(0.34 \sin 10.2°)$$
$$- (622.8)(0.85 \sin 10.2°) = 0$$

We now have three equations in three unknowns:

$$0.342\ F_m - F_{aX} = 0$$
$$0.94\ F_m - F_{aY} = -525.64$$
$$0.0423\ F_{aY} - 0.0426\ F_{aX} = 87.9$$

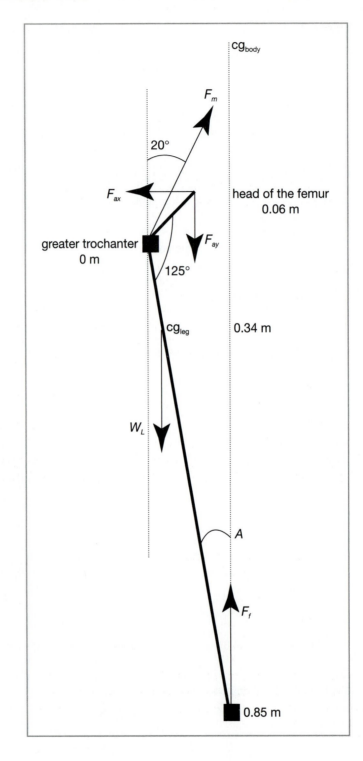

These equations may be solved either by algebra or by using a scientific calculator to yield

$$F_m = 4371 \text{ N} \quad (983 \text{ lb})$$
$$F_{aX} = 1495 \text{ N}$$
$$F_{aY} = 3584 \text{ N}$$

Using the Pythagorean theorem, we get

$$F_a = 3.88 \times 10^3 \text{ N} \quad (873 \text{ lb})$$

Notice that the force compressing the head of the femur is almost 6 times the person's body weight. This is because the hip abductor muscle passes so close to the joint between the femur and the acetabulum that its lever arm is very small, and thus the tension in this muscle is very large (983 lb). This then results in a large force (873 lb) of compression at the joint.

EXAMPLE 2.14

Now assume that the same person has an injured right hip. The large amount of force between the acetabulum and the head of the femur will produce a great deal of pain. To lessen the pain, the person will automatically lean to the right when walking, therefore limping. The person has shifted his weight so that the right leg is closer to the vertical (assume 6° from the vertical). It may reasonably be assumed that under these circumstances, the hip abductor muscles pull at an angle of 5° from the vertical.

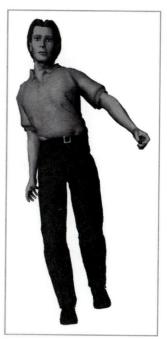

1. Determine the magnitude of the contact (or reaction) force acting between the acetabulum and the femur.
2. What has been accomplished by the shifting?

Solution

The analysis of this problem is very similar to that of Example 2.13. The only changes are the two angles.

ANSWERS:

$$F_{aY} = 1197 \text{ N}$$
$$F_m = 673.7 \text{ N} \ (151 \text{ lb})$$
$$F_{aX} = 58.71 \text{ N}$$
$$F_a = 1.2 \times 10^3 \text{ N} \ (269 \text{ lb})$$

Notice that the force compression the acetabulum-femur joint has been reduced from 873 lb to 269 lb by the person's leaning to the right side when walking. This explains why someone who has an injured hip will lean toward the injured side and limp as a result.

EXAMPLE 2.15

In Example 2.14, we showed that such a person would auto-matically shift his or her weight to the right to lessen the force between the femur and the acetabulum. Unfortunately, this shift causes the person to limp and also causes stress both on the top of the femur and on the person's back. A better alter-native is for the injured person to use a cane on the contralat-eral (i.e., left) side. In this problem, you are asked to analyze the effect of using a cane on the force between the femur and the acetabulum. Consider a person whose weight is support-ed by the right leg and by a vertical cane on the left side. The end of the cane rests on the floor at a point 30 cm to the left of the person's center of gravity and supports one sixth of the person's weight. The angle between the hip abductor muscles and the vertical is 20°.

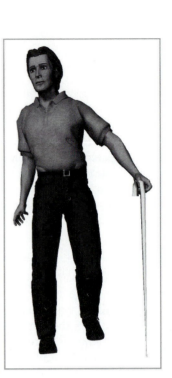

1. How much force is acting between the floor and the person's right foot?
2. What is the distance, measured horizontally, between the person's right foot and the person's cg?
3. How much force is acting between the person's femur and acetabulum?
4. What has been accomplished by the use of the cane?

Solution

To deal with the cane, we must realize that there are two basic differences between supporting the body only by the right leg and supporting the body by the right leg and a cane on the left side. One of these is the amount of force exerted by the floor on the right foot. In the former case, the right foot supports the entire body weight; in the latter case, the weight is shared by the cane and the right foot. The other difference is the location of the cg of the body. In the for-mer case, the cg must be directly over the right foot, whereas in the latter case, the cg is locat-ed somewhere between the cane and the right foot.

1. Once again, the object of interest is the right leg.
2. The forces acting on the right leg are the same except the force of the floor F_f, which is no longer equal to 622.75 N.
3. The FBD of the right leg will basically be the same as before (only the angles are differ-ent).
4. Notice that there are now five unknowns: F_f is not known, and the angle between the leg and the vertical is unknown; they must be calculated.
5. To calculate F_f, we must consider a FBD of the entire body. The body is represented by the rectangle in the figure. The only forces acting will be the weight of the body (622.75 N), the force of the floor on the cane (103.79 N), and the force of the floor on the right foot (F_f).

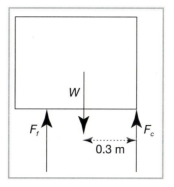

6.
$$\uparrow + \sum F_Y = 0$$

$$F_f + F_c - W = 0$$

$$F_f + 103.79 - 622.75 = 0$$

$$F_f = 519 \text{ N}$$

7. We do not know the horizontal distance between the right foot and the cg of the body. This may be determined from the torque equation. Let the axis of rotation be at the center of gravity:

$$\sum \Gamma_0 = 0$$

$$(F_f)(X) - (F_c)(0.3) = 0$$

$$(519)(X) - (103.79)(0.3) = 0$$

$$X = 0.06 \text{ m}$$

Since the distance from the greater trochanter to the cg of the body is 15 cm and the distance from the right foot to the cg of the body is 6 cm, the distance from the greater trochanter to the right foot is 9 cm.

8. The diagram on page 116 shows the FBD.
9. The angle A may be calculated by considering the right triangle whose hypotenuse is the 0.85-m-long diagonal line: $A = 6.1°$.
10. We now have similar information as given in Example 2.14. The same three basic equations are written and solved for the unknowns:

$$F_{aY} = 1183 \text{ N}$$
$$F_m = 810 \text{ N} \quad (182 \text{ lb})$$
$$F_{aX} = 277 \text{ N}$$
$$F_a = 1.22 \times 10^3 \text{ N} \ (273 \text{ lb})$$

Thus we see that if a 140-lb man supports his weight on one leg without leaning, the force of compression at the head of the femur will be 873 lb. If he purposely leans to the right, the force will be reduced to 269 lb, but he may injure his femur and/or his back. If he correctly uses a cane, the force will be 273 lb; however, there will be much less chance of damage to the femur or the back.

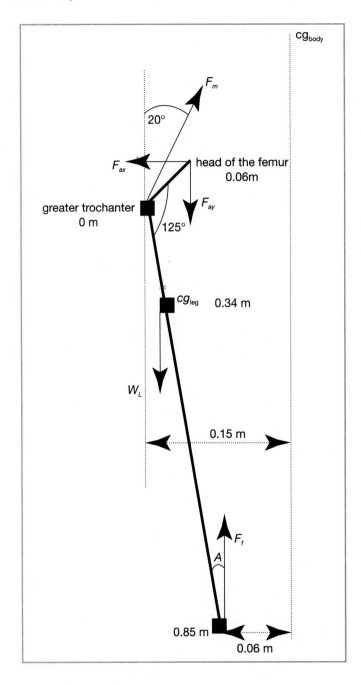

PROBLEM SET 6

Note: You may need to use the anatomy material in Appendix 7 for additional data on some problems.

6.1. Explain why a pregnant woman typically leans backward when standing.

6.2. Why might the practice of carrying a heavy load on one's head rather than in one's arms be better for the person's back.

6.3. A person tries to unscrew the top of a jar by applying a force of 3 lb tangent to the rim of the screw-top, which is 1 inch in diameter. How much torque is being applied? The person now uses a kitchen tool that is 4 inches long. Assuming that the same force is applied, how much torque results? (0.125 ft lb, 1 ft lb)

6.4. A person is holding 5 kg of potatoes out to the side at arm's length (85 cm). How much torque does the load produce on the arm? (41.65 mN)

6.5. A box rests on a 10-ft, 7-lb board that is supported by two bathroom scales separated by 7 ft. The scale on the left is located 1.5 ft from the left end and reads 45 lb. The scale on the right reads 25 lb.
   a. Determine the weight of the box. (63 lb)
   b. Locate the center of mass of the box. (3.9 ft from the left end)

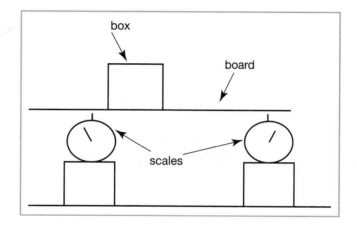

6.6. Consider an 846-g baseball bat that is 42 inches long. When a 250-g mass is taped onto the bat at a point 38 inches up from the small end, the balance point of the bat-mass combination is found to be located 28 inches up from the small end of the bat. Determine the location of the cg of the baseball bat itself. (25 inches from the small end of the bat)

6.7. A skeleton, holding up the end of an 18-ft-long 84-lb nonuniform log, applies a force F_1 at the right end of the log, as shown in the figure. The cg of the log is located 3 ft from the right end. The angle between the log and the horizontal is 30°.
   a. Calculate the magnitude of F_1. (70 lb)
   b. A friend, seeing the skeleton's efforts, comes over to help by lifting straight up at the other end of the log. Does this help the skeleton or not? Explain.

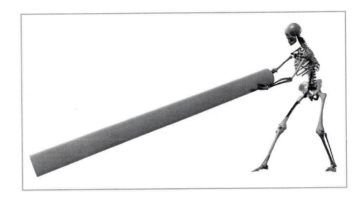

6.8. The angle between the ladder and the wall is 34°.
      The 115-lb painter is standing on the 12-ft, 30-lb
      ladder at a point 3 ft from the upper end. Assume
      that there is no friction between the ladder and the
      wall (this means that the only force at the upper end
      of the ladder will be directed perpendicular to the
      wall). There is friction at the lower end, between the
      ladder and the ground, so there will be both a hori-
      zontal force and a vertical force there.

      a. Determine the magnitude of the force acting on
         the ladder at the upper end. (63.2 lb)
      b. Determine the magnitude of the force acting on
         the ladder at the lower end. (158 lb)
      c. Determine the direction of the force acting on the
         ladder at the lower end. (66° to the right and
         above the horizontal)

6.9. The crane has as its main structural component, a
      30-ft-long, 2300-kg boom. The angle between the
      cable and the horizontal is 32°, and the angle
      between the boom and the vertical is 40°. It is sup-
      porting a 5300-kg load by a cable that is attached to
      the end of the boom.

      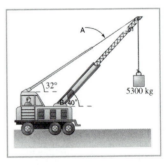

      a. Determine the amount of force in the cable at
         point A. (353 kN)
      b. Determine the amount of force of compression at point B, the bottom end of the
         boom. (397 kN)
      c. Find the direction of the force that is exerted on the boom at point B. (directed
         up and to the right at 41° above the horizontal)

6.10. A 23-kg sign is suspended from the end of a 7-kg pole that is 4 m long. The pole is
      partially supported by a cable that is attached at a point 3 m up along the pole from
      the socket. The angle between the pole and the vertical wall is 60°, and the angle
      between the cable and the wall is 70°. Determine the magnitude and direction of the

force exerted by the wall socket on the bottom end of the pole. (401 N, 23.5° CCW from the horizontal)

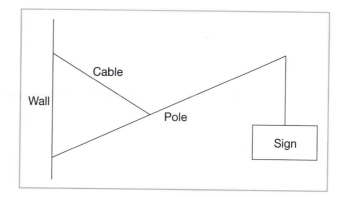

6.11. A woman who is carrying two 35-lb garbage cans out to the road notices pain in her shoulders. This problem deals with that pain. Assume that one can is held by each hand and, to prevent getting her clothes dirty, she holds the cans to each side, away from her body. Under these conditions, the muscle that is under the greatest stress is the deltoid. The deltoid (represented by the arrow in the figure) inserts

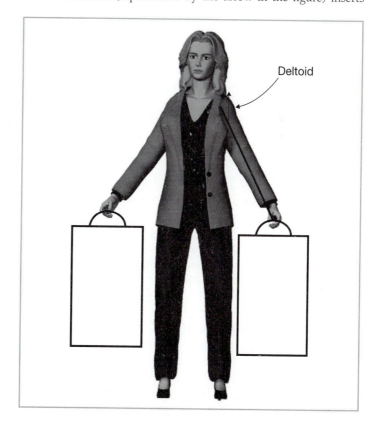

onto the humerus at a point 5 inches below the shoulder and makes an angle of 15° with the axis of the arm. Her arm (represented by the heavy dark line in the figure) weighs 7.2 lb, is 24 inches long, and makes an angle of 20° with the vertical. How much force is exerted on the upper end of the humerus? (201 lb)

6.12.   A man is standing erect. We are going to concentrate on his left arm. The upper arm is vertical, and the lower arm makes an angle of 30° with the vertical. The figure shows his elbow.

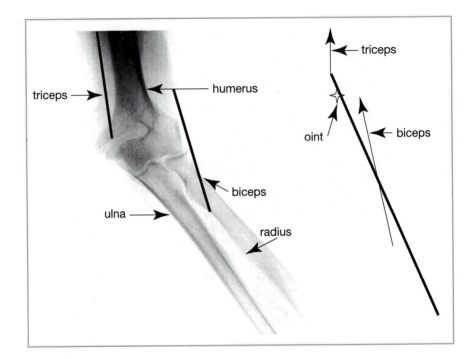

The following data are given:

- Distance from humerus/radius/ulna joint (elbow) to hand: 15 inches
- Biceps brachii muscle: inserts onto the radius at a point 2 inches in front of the ulna-humerus joint. The angle between the biceps tendon and the forearm is 10°. This muscle plays a major role in causing flexion of the forearm, that is, rotation toward the upper arm.
- Triceps brachii muscle: inserts onto the ulna at a point 1 inch behind the ulna/radius/humerus joint. This muscle plays a major role in causing extension of the forearm, that is, rotation away from the upper arm.
- Body weight: 190 lb
    a. In this part, the man is holding a 10-lb weight in his hand. Determine the magnitude and direction of the force exerted by the upper arm on the forearm. (253 lb, 69° below the horizontal toward the hand)
    b. In this part, the man is using a vertical cane in his left hand to help him stand. The cane supports one sixth of his body weight. Determine the magnitude

and direction of the force exerted by the upper arm on the forearm. (441 lb, straight down)

6.13. Consider a 120-lb woman who is using an exercise machine that is designed to stress the hamstring muscles. These muscles originate at the hip, pass along the femur, and then terminate in a tendon that inserts onto the tibia at a point located 4 cm from the knee. Her feet are positioned so that a bar, to which the cable is attached, passes behind the ankle. When the knees are flexed, the weights are pulled up. (*Note*: The diagram on the right is not intended to be a complete FBD.)

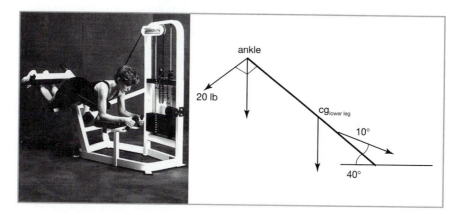

Specific data relating to this problem:

- Angle between the hamstring tendon and axis of the tibia: 10°
- Angle between the axis of lower leg and the horizontal: 40°
- Angle between the force of the pad behind the ankle and the axis of the lower leg: 90°
- Amount of weight on the machine: 20 lb

Determine the magnitude and direction of the force that the femur exerts on the tibia. (680 lb directed up and backward at 31° above the horizontal)

6.14. When the body is in an erect posture, its cg lies on a line that falls 1.25 inches in front of the ankle joint. The calf muscle (the Achilles tendon) attaches at the ankle 1.75 inches in back of the joint and passes up at an angle of 83° measured from the horizontal (as shown on p. 122).

    a. Calculate the force T in this muscle for a 150-lb person standing erect. Remember that each leg supports only half of the person's weight. (54 lb)

    b. Determine the magnitude and direction of the reaction force R exerted at the ankle joint. (129 lb at 87° below the horizontal)

6.15. In Example 2.11, we dealt with the problem presented by a 200-lb person who is squatting while holding a 300-lb weight. In that example, we were concerned with the forces that were acting at the person's knee. In this problem, we are concerned with the forces that act at the person's ankle. It is known that the Achilles tendon (T) inserts onto the heel at a point (A) that is 7.3 inches from the ball of the foot (C) and that the tendon pulls perpendicular to the axis of the foot (ABC).

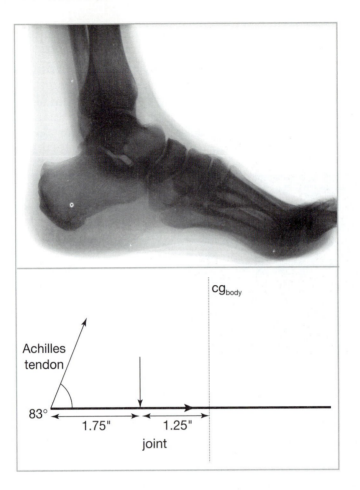

The line of action of the force exerted by the tibia on the talus intersects the axis of the foot at point *B*, which is 3.8 inches from the ball of the foot. The axis of the foot (*ABC*) makes a 50° angle with the horizontal. Find the magnitude and the direction of the force that the tibia exerts on the foot. (*Note*: The weight of the foot may be ignored in this problem.) (386 lb, directed at 20° CW from straight down)

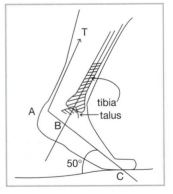

6.16. A 120-lb person is lifting 20 lb, using a quadriceps exercise machine. When his quadriceps contract, his ankles push on the bar, thus lifting the weights. The force exerted by the bar is perpendicular to the axis of his lower leg. The quadriceps tendon makes an angle of 5° with the axis of the lower leg. Its LA (assuming that the axis of rotation is at the femur-tibia contact) is 6.5 cm. (*Note*: The sketch on the right on page 123 is not intended to be complete.) Consider only one of his legs (the angle between the leg and the vertical is 50°) and

a. Calculate the force exerted by his quadriceps tendon on his tibia. (322 N (= 72 lb)).

b. Determine the magnitude and direction of the force exerted by his femur on the top of his tibia. (304 N (= 68 lb)) directed down and to the right at 32° below the horizontal)

6.17. The figure shows a 70-kg person using a back extension machine on which there is a load of 120 lb. The angle formed by his back and the horizontal is 40°. He is pushing (perpendicular to his spinal column) against a pad that is 50 cm from L5-S1 (measured along the spinal column). The length of the spinal column (from L5-S1 to the neck) is 60 cm. The erector spinae muscle makes an angle of 12° with the spinal column and inserts onto the spinal column at a point 0.67 of the distance up from L5-S1. Assume that the cg of his arms is located halfway up along the spinal column. Determine the amount of compression at L5-S1. (2436 N (= 548 lb))

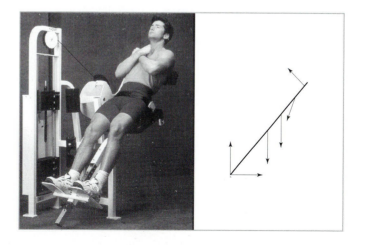

MECHANICAL ADVANTAGE

As we have seen in several examples, the body makes use of muscles or tendons acting with a relatively short lever arm around a joint to produce rotation or to counter another force that is usually associated with a much larger lever arm. For example, when analyzing the outstretched arm supporting a weight, we saw that the deltoid muscle had a very short lever arm but the weight in the person's hand had a very large lever arm. As a consequence, the force exerted by the deltoid was many times larger than the force associated with the weight being supported. If we define the force associated with the deltoid muscle as the *cause force* and the weight of the supported object as the *effect force*, we may write:

$$F_{cause} > F_{effect} \qquad \text{because} \qquad LA_{cause} < LA_{effect}$$

These may be rewritten as:

$$\frac{F_{cause}}{F_{effect}} > 1 \qquad \text{and} \qquad \frac{LA_{effect}}{LA_{cause}} > 1$$

From a consideration of the two torques, we know that:

$$\Gamma_{effect} - \Gamma_{cause} = 0$$
$$\Gamma_{effect} = \Gamma_{cause}$$
$$F_{effect}LA_{effect} = F_{cause}LA_{cause}$$
$$\frac{LA_{effect}}{LA_{cause}} = \frac{F_{cause}}{F_{effect}}$$

This proportionality represents the relationships that we described above. It defines what is commonly called the *mechanical advantage* (MA):

$$MA = \frac{F_{cause}}{F_{effect}} = \frac{L_{effect}}{L_{cause}} = \frac{D_{effect}}{D_{cause}}$$

D_{cause} is the distance through which F_{cause} moves, and D_{effect} is the distance through which F_{effect} moves. We discuss these distances in the following example.

EXAMPLE 2.16

The deltoid muscle causes abduction of the arm. When it contracts, the arm rotates away from the body. Look back at the picture in Problem 6.11. In this problem, the cause is represented by the deltoid muscle, and the effect is represented by the garbage can. D_{cause} is the distance through which the deltoid muscle moves (contracts), and D_{effect} is the distance through which the garbage can moves:

$$MA_{deltoid} = \frac{F_{garbage\ can}}{F_{deltoid}} = \frac{LA_{deltoid}}{LA_{garbage\ can}} = \frac{D_{deltoid}}{D_{garbage\ can}}$$

Since the distance that the garbage can moves is much larger than the distance that the deltoid muscle moves, we see that the *MA* of the deltoid muscle is much less than 1. This is just one example of the many skeletal muscles. The result for each of them would be the same: *MA* less than 1.

The human body is rather weak in relation to many of the tasks that we try to carry out. For example, consider lifting heavy objects such as rocks, blocks of stone, or cars. Normally, we use tools and simple machines to accomplish what we could not accomplish otherwise. For example, a lever allows me to exert a much larger force than I could exert without it. I certainly cannot pull a nail out of a board with my hand. However, if I use a claw hammer so that my force (the input force) is associated with a large lever arm and the axis of rotation is close to the nail so that the nail has a short lever arm, the small force produced by my body will result in a very large force (the output force) being applied to the nail. It may then be pulled out. In this case, the *MA* would be much larger than 1. A relatively small input force results in a very large output force. This is characteristic of almost all simple machines, such as the lever, the inclined plane, and the screw. It also applies to a car jack that has a very large *MA*. A relatively weak person can actually lift a car up into the air. By our invention and extensive use of tools (that have a built-in large *MA*), humans have compensated for fundamentally low *MA* bodies. This inherent low *MA* seems to be a strangely inefficient design. The good news has to do with the distances that were mentioned, but not discussed, above. Recall that the *MA* may be defined as

$$MA = \frac{D_{\text{cause}}}{D_{\text{effect}}}$$

Thus a small *MA,* such as that associated with skeletal muscle, means that a small displacement of the cause force will result in a large displacement of the effect force. If the quadriceps muscle contracts only a few millimeters, the person's foot may move through a meter (1000 millimeters). Similar examples are the biceps muscle (elbow), and the masseter muscle (jaw). In all of these cases, a small amount of movement within the body results in much more motion at its extremities. This provides for much faster motion of arms, legs, and head. Thus we seem to have evolved favoring inherent quick response over strength, while using tools to compensate for the lack of strength.

3

HEAT AND ENERGY

INTRODUCTION TO HEAT AND ENERGY

Note: In the following discussion, heat is referred to as *flowing, being generated, produced, gained,* and *lost.* These terms refer to the commonly held concept of heat and do not reflect the nature of heat as dealt with by the physical sciences. The correct model will be discussed later. We humans share a significant problem with a large number of other creatures. For us, as for all other warm-blooded animals, keeping our body temperature within a fairly narrow range is a major issue of survival. A reasonable value of normal internal body temperature for an adult is 37°C (99°F). If your internal temperature is not very close to this value, you may be in danger of dying.

For example, as your core temperature drops to 35°C (95°F), intense shivering and an inability to coordinate muscular activity are experienced. At about 32°C (90°F), your central nervous system behaves erratically, you may have trouble speaking, and you may hallucinate and experience amnesia. At 30°C (86°F), muscles become rigid, the skin turns blue, and then you may pass into a state of stupor. Below 30°C (86°F), you lose consciousness and approach death as the temperature falls to 26°C (78°F). Notice that a person may die at a temperature that is far above freezing. The effect of low core temperature on the body is called *hypothermia* and is one of the major dangers faced by people who are exposed to an ambient temperature (temperature of the environment) that is less than normal body temperature.

Elevated body temperature is caused by fever or *hyperthermia*. Fever is associated with an increase in internal temperature as caused by the body itself. This is usually due to infection and may be part of the body's defense mechanism. *Hyperthermia* is associated with the body's interaction with its environment. For example, the body may be producing heat more quickly than the heat is lost, or the ambient conditions may be such that heat may not be removed from the body to compensate for even a normal rate of heat production. An extreme example of hyperthermia is heatstroke. This condition results from a failure of the body's automatic temperature regulation mechanisms (to be discussed later) to properly react to a high ambient temperature.

The normal ambient temperature for most people in the world is less than 37°C. Therefore most people live in a situation in which they are surrounded by an environment that is at a lower temperature than the body. We know that under normal conditions, heat flows from a higher-temperature place to a lower-temperature place. Therefore all of these people are continuously losing heat to their environment. A loss of heat usually results in a decrease in temperature, so hypothermia looms for all of these people. Yet hypothermia is rarely encountered. This happy situation results from a set of functions that are automatically triggered in a warm-blooded animal's body. These various functions serve to generate sufficient heat within the body to compensate for the loss to the environment and to decrease the flow of heat out from the body. Included in these processes are the rate and paths of blood circulation, metabolic rate, shivering, and "gooseflesh." It is only when these processes fail that hypothermia results.

A mammal may have as much as a 30°C difference in temperature between the core of the body and its extremities. Even in a human, the temperature of the blood in the radial artery of the arm has been measured as low as 21.5°C when the core temperature was 37°C. (Lenihan, 1975, p. 134).

We shall see that the temperature of an object may increase if the amount of heat in that object increases. There are several different phenomena that would cause the amount of heat in a body to increase.

Heat Exchange with the Environment

As was mentioned earlier, heat naturally flows from a higher-temperature region to a lower-temperature region. Therefore heat will flow into your body from the environment if the temperature of the environment is higher than the temperature of your body ($T_{environment} > 99°F$). Before the advent of air conditioning, people who lived in climates that typically experienced such high temperatures had to severely restrict their lifestyles, for example, by working early in the morning, taking a rest during the early afternoon, and then working again in the late afternoon and evening.

Internal Heat Generation

Heat is generated within your body when muscles contract. The contraction of your heart, a collection of muscle whose main function is to make blood circulate, is a major source of heat within the body. Heat is also generated as a by-product in many chemical reactions. Your liver, which produces many of the chemicals (e.g., cholesterol) that are used by the body, is also a major source of heat. Neurons produce heat as a by-product of the processes associated with the ionic transport that characterize their activity. Thus the brain (a massive collection of neurons) is a center of heat production in the body.

If heat is produced in a certain region of the body (brain, liver, heart, etc.), then that heat must be removed, or the temperature of that region will probably increase. The body has several ways of effecting internal heat transfer, among which the most effective is the circulatory system. Not only must heat be moved around within the body, it must also be transferred away from the body. This is accomplished by several processes. The most obvious of these are sweating, panting, and making use of air currents in the environment.

Heat is a widely accepted form of therapy for the human body. Localized heat has two major effects:

1. The metabolic rate in the affected tissue increases.
2. There is an increase in blood circulation within the region.

The former makes it easier for cells to multiply, perhaps replacing damaged tissue, and the latter both brings oxygen and nutrients for the increased cell growth and removes debris from the damaged tissue.

The localized removal of heat from the body is also a form of therapy. This intervention will initially cause a marked decrease in neural activity (local anesthesia) and will ultimately cause tissue damage (cryosurgery).

To understand how the body manages to regulate its temperature so well, we must deal with two distinct physical concepts: temperature and heat.

Although temperature is the more familiar term, it is heat that is the more basic concept in physics. We have used the words, "heat" and "energy" many times in the past few pages without defining them. As it turns out, arriving at acceptable definitions is very difficult.

The major reason why it is so difficult to explain "heat," "energy," and "temperature" is that they cannot be directly observed. Explaining a rock is not difficult; just pick out some examples and put them out for observation and experimentation. However, the words that we are trying to explain cannot be demonstrated in this way. We are limited to demonstrating heat, energy, and temperature (along with many other words that are commonly used in physics, such as electricity, sound, magnetism, and gravity) by dealing with their effects on things that can be directly observed. For example, when an object is heated, it may:

1. change shape or size
2. change physical state (solid, liquid, gas)
3. change chemical state (e.g., combustion)
4. change optically (e.g., a "mood ring")
5. change electrically (e.g., a semiconductor)
6. change magnetically (e.g., an iron magnet)

Any explanation (or model) that is suggested must explain all of these observations (and many more). Since heat, energy, and temperature must be explained by using indirect observations, it should not be surprising that several explanations have been offered.

Among the earliest modern writings about heat, there is the work of Isaac Newton. In one of his books (*Opticks*, 1704–1730), he used "Queries" to denote concepts that he felt were useful in understanding the physics of the world but that could not be either direct-ly observed or proven by the use of reasoning and mathematics. One of these queries (Query 5) was ". . . and Light upon Bodies for heating them and putting their parts into a vibrating motion wherein heat consists?" In this passage, Newton is suggesting that heat is really the vibratory motion of the particles of which an object is composed. We shall see that Newton's idea was right on the money.

Newton's answer to the question "What is heat?" was not an "acceptable explanation." So what is an acceptable explanation? Quite often, when someone learns that I am a physi-cist, I may be asked to explain something such as "Why isn't there any antigravity?" or "Is time travel possible?" The issue that I try to deal with is "What will the person accept as an explanation?" If I refer to mathematics or theories such as quantum mechanics or rela-tivity, the person will probably feel that I am snowing him or her and that I really don't understand the issue or I could make it clearer. So what is an acceptable explanation? I suggest that an acceptable explanation is one that refers to ideas with which the person is already comfortable. The ideas might not be completely understood, but if they are accepted, then so too will be the explanation.

Newton's explanation of heat was based on the idea that real bodies were composed of infinitesimal particles that were in random motion. We can understand why this expla-nation was put into the form of a "Query" since it occurred approximately 100 years before the concept of atoms was seriously considered by the scientific community. Some other, more acceptable explanation was required, and one was developed.

There is another issue that must be addressed in evaluating a scientific explanation. That is the limitation imposed by language. This question is usually associated with Noam Chomsky, a linguist who works at MIT. He has taught that a person's ability to formulate concepts is limited by the person's language. The more limited the language, the more lim-ited the concepts. Consider the question "What is heat?" In this sentence, the word "heat" is a noun. Most of us have learned that a noun is the name of a person, place, or thing. Heat is certainly not a person or a place, and so it must be a thing. Therefore when we say the word "heat," we have already assumed that the word refers to some sort of thing or stuff. Thus one can gain or lose heat, heat may go out through a window, heat may be pro-duced or consumed. We think of it as we do other types of stuff, such as water or air. Newton's idea that heat is related to the random motion of the particles that make up an object does not satisfy this preconception. However, an explanation of heat that was more acceptable was developed: the caloric theory.

By the late 1700s, the idea of **imponderable fluids** had become popular in natural philosophy, as physics was then known. *Imponderable* meant without weight. Thus peo-ple thought that a way of explaining various properties of an object was to assume that within the object, perhaps several different kinds of fluid were circulating. These fluids could not be directly observed, but one could tell that they were present by observing the properties of the object. This type of explanation was widely accepted, not only by scientists but also by the culture at large. As late as 1820, a majority of the scientists accepted the imponderable fluid explanation of heat. (Brush, 1964, p. 488.) In 1858, Oliver Wendell Holmes, an eminent U.S. intellectual nonscientist, wrote, in *The Autocrat of the Breakfast Table,* "It is the imponderables—heat, electricity, love—that rule the world."

For example, one could say that as something became warmer or colder, it either gained or lost "something." You might say that heat was "going out" through a thin coat or through a poorly insulated wall or window. These comments imply that heat is some sort of stuff that can be lost or gained or can go in or out. Since this stuff did not have an observable shape, it was imagined as a fluid. By a similar argument, if after combing your hair, you noticed that the comb would attract small bits of paper or dust, you might say that something had been put onto the comb that gave it this ability. Whatever it was that the comb gained was also imagined to be a fluid (later called electricity). Exactly where these fluids came from or how they were passed from one object to another were open to speculation.

The particular imponderable fluid associated with heat was called **caloric.** The caloric model was developed extensively during the eighteenth and early nineteenth centuries, especially by chemists such as Joseph Black (1728–1799). He seems to have been the first person to distinguish between heat (the quantity of heat) and temperature (the intensity of heat). In 1760, he introduced the idea that the quantity of heat could be measured by the change in temperature of a specific quantity of matter. He defined an amount of heat, the **British thermal unit (Btu),** as the amount of heat that would increase the temperature of one pound of water by 1°F. This meant that he could calculate the amount of heat that was involved in an experiment by measuring the change in temperature of one pound of water. Because of this contribution, he is given much of the credit for putting the study of heat on a quantitative rather than a qualitative basis. In one of his experiments, he measured the amount of heat that was required to melt a certain amount of ice. He did this by placing equal amounts of ice and water on a stove. He assumed that heat from the hot stove would go into the ice and into the water at the same rates, and therefore he could compare the amounts of heat passing into each substance by comparing the amounts of time required for the transfers. He found that it took three times longer for the pound of ice to melt than for the pound of water to warm up from 32°F to 212°F. On this basis, he determined that the amount of heat required to melt a given amount of ice was 540 times larger than the amount of heat required to change the temperature of the same amount of water by 1°F. That is, ice requires 540 Btu/lb to melt. These data led to the concept of latent heats, to be discussed on page 176–181.

The **caloric theory** satisfied many of the expectations for an acceptable explanation. It referred to an already accepted idea—fluids—and it referred to a sort of stuff, consistent with the use of a noun. The use of the caloric model reached its highest point in the work of Sadi Carnot (1796–1831). In 1824, he published a theoretical explanation of the efficiency of the steam engine. His theory was built on the caloric model and was so important that it has become one of the building blocks of what is called Classical Physics.

However, there were some who questioned the caloric model. The first of these was a dynamic man named Benjamin Thompson (1753–1814). He lived in pre-Revolutionary America and was a Tory, that is, he supported and fought on the British side in the Revolutionary War. Because of this, he left at the close of the war and became a consulting military engineer. His significant work with caloric occurred while he was in the employ of the ruler of Bavaria, a principality that later became part of Germany. Thompson was involved in the manufacture of artillery. In those days, artillery was made by casting the cannon from molten metal. Unfortunately, the process of casting usually results in rough surfaces and thus inefficient cannon. To achieve smoother inner surfaces, the casting was bored out. During the boring, the casting got very hot and often cracked. This heating up could be alleviated by letting water run onto the casting during the boring process. The water would evaporate, and the casting would not overheat to the point of cracking. This was explained as follows: The casting got hot because it gained caloric, the water absorbed some caloric from the casting, and as a result, some of the water evaporated. The water that evaporated took caloric away from the casting. Thompson seems to have been the first person to wonder (and write about) "Where did the caloric that the casting was picking up come from?" It was commonly accepted that caloric always went from something hot (which, since it lost caloric, became cooler) to something cold (which, since it gained caloric, became hotter). But Thompson noticed that in the boring process, nothing was becoming cooler. The boring tool became hotter, the casting became hotter, and the water became hot enough to evaporate. Thompson conjectured that the caloric came from the horses that were making the boring tool go around as it wore down the roughness on the casting. Of course, the horses were not getting cooler, they were also getting hotter. (*Note:* Thompson was made a nobleman by the ruler of Bavaria and took the title Count Rumford in honor of the town in New Hampshire where he was born. He later used his accumulated wealth to endow the Royal Institution, which became one of the leading centers of scientific research in England.)

The next step in the demise of the caloric theory is credited to two investigators: the German physician Julius Robert Mayer (1814–1878) and the English brewer James Prescott Joule (1818–1889). Mayer was employed as a physician for Europeans who were working in the East Indies. He noticed that their venous blood, which in Europe had been colored dark red, was in the tropics colored as bright a red as the arterial blood. On the basis of his studies of the work of the French chemist Lavoisier, Mayer suggested that the amount of oxygen in the venous blood of a person in the tropics is higher than that in the blood of a person in a temperate climate such as Europe. This made sense in that, since it is hotter in the tropics, there is less need for a high metabolism and thus less consumption of oxygen. He went further and suggested that there is an exact balance of force (which we now call "energy") in the body. The energy released by the food is balanced by the lost body heat and the work done by the body. Mayer wrote in 1842, "Once in existence, force (energy) cannot be annihilated—it can only change its form." Unfortunately, Mayer's work did not receive the attention that it warranted, and he did not get the credit he deserved.

Joule was one of a fascinating group of people who lived in the seventeenth and eighteenth centuries. They were "gentlemen scientists." Benjamin Franklin was also a member of this group. They were people who were wealthy enough from other pursuits to dabble in science as a hobby or avocation. Joule had inherited one of the largest breweries in England and spent a great deal of time doing experiments in what we now call thermal physics. His most significant experiment (in 1843) showed that there is a strict numerical relationship between the amount of temperature increase of a vat of water and the motion of a falling weight that drove a set of paddles stirring the water. He defined a unit of work to be the amount required to raise one pound through a distance of one foot. He found, through a series of many experiments, that one unit of heat (Btu) corresponded to a specific number of units of work (lb ft). The number is now accepted as 1 Btu = 778 lb ft.

In 1847, he delivered a lecture, later published in a Manchester weekly paper as "Matter, Living Force and Heat," in which he stated:

> Living force (*vis viva*) is one of the most important qualities with which matter can be endowed and, as such, it would be absurd to suppose that it can be destroyed. . . . Experiment has shown that wherever living force is apparently destroyed, whether by percussion, friction, or any similar means, an exact amount of heat is restored. The converse is also true, namely, that heat cannot be lessened or absorbed without the production of living force or its equivalent attraction through space. . . . Heat, living force and attraction through space (to which I might also add light, were it consistent with the scope of the present lecture) are mutually convertible. **In these conversions nothing is ever lost** (Singer, 1959, p. 376).

The last sentence is particularly important. In it, **Joule suggests that although living force can be converted from one form to another, the total amount of living force cannot change; it cannot be created or destroyed. What Joule called "living force" is now called "energy."** That term was coined by William Thomson (1824–1907), who is commonly referred to as Lord Kelvin, in 1852.

This idea that the total amount of energy cannot change is basic to our understanding of the world. I mean that term "world" in the fullest sense. Whether you study astronomy, biology, geology, or economics, this basic concept will play an important role. It is a difficult concept, however, because energy cannot be seen or otherwise directly detected. The flow of energy from one object to another and conversion from one form into another can be dealt with by using the concept that was invented by Thompson, Mayer and Joule. It is called **Conservation of Energy** or the **First Law of Thermodynamics.** It may be simply stated in words as follows:

Energy cannot be created or destroyed. It may only change from one form to another.

Mathematically, it may be stated as follows:

$$E_1 = E_2$$

Thus when we eat some food (this represents energy added to the body, that is, us), we may do some work (giving energy to something else, such as lifting a weight), we may increase our internal energy (store some of the energy as fat) and/or we may produce some heat.

EXAMPLE 3.1

In the following example, I am trying to illustrate the idea of conservation; I am not implying that energy is a fluid such as water. We can try to picture what is going on by imagining a house and considering the plumbing system.

1. Water may enter the house from the water main or from a well.
2. Some of the water may be stored in a hot water tank.
3. Some of the water may leave the house via the sewer or septic system.
4. Some water may leave the house because people or pets drank it and then walked out.
5. Some water may leave the house because water evaporates.
6. Some water may enter the house because of humid air coming in.

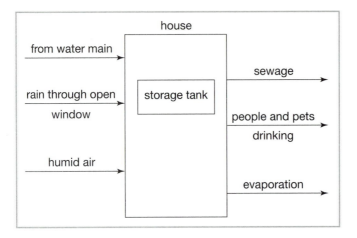

7. Some water might even enter the house because it is raining and someone left a window open.

It is important to understand that since water cannot be created or destroyed (it can only change form, for example, from liquid to vapor), we must be able to specifically account for any change in the amount of water in the house. We could write an equation (in which we will use W to represent an amount of water):

$$W_{\text{in from water main}} + W_{\text{in with humid air}} + W_{\text{in through open window}}$$
$$= W_{\text{stored}} + W_{\text{out to sewer}} + W_{\text{out via evaporation}} + W_{\text{out via people and pets}}$$

Each of these terms is defined as a positive number but could become negative either by moving it to the opposite side of the equation or by changing the preposition in its identification. For example,

$$W_{\text{in from water main}} + W_{\text{in with humid air}} - W_{\text{out through open window}}$$
$$= W_{\text{stored}} - W_{\text{in from sewer}} + W_{\text{out via evaporation}} - W_{\text{in via people and pets}}$$

or

$$W_{\text{in from water main}} + W_{\text{in with humid air}} - W_{\text{out via evaporation}}$$
$$= W_{\text{stored}} - W_{\text{in from sewer}} - W_{\text{in via people and pets}} + W_{\text{out through open window}}$$

EXAMPLE 3.2

Imagine that you live in a location where you are not allowed to dispose of sewage by letting it go into the ground. It is necessary that all sewage be collected in an underground tank and then be periodically pumped out into a truck and taken away. Suppose that you get a bill from the local water company in which you are charged for using 400 ft^3 of water in a given month. During the same month, you are charged for the removal of 500 ft^3 by the waste disposal company. How could this be explained?

Since more water is being removed than came in from the water main and we firmly believe that water cannot just appear, something is wrong. Perhaps someone is bringing jugs of water home from work and dumping them into the sink. Or perhaps your neighbor has connected his sewer line to your underground tank so that you are paying for the removal of his sewage. It is more likely that there is an underground leak into your storage tank from a spring.

The important idea for this example is that water cannot just appear (or disappear), it must come from somewhere. In later problems, we shall say that energy cannot just appear or disappear, it must come from somewhere (or go somewhere). While dealing with something tangible (water in Example 3.1 or money in a household budget), it is quite reasonable to believe that it (the tangible material) cannot be created or destroyed (unless you burn the money or have a printing press in the cellar). It can only move around, perhaps changing forms. It is much more difficult to accept this concept when it is applied to something such as energy that is completely intangible. Common sense leads us to "know" that heat can be made or lost, that light or sound can appear and disappear, and so on. This dichotomy between common sense and a more fundamental level of explanation is basic to epistemology, the part of classical philosophy that deals with the nature of knowledge. We can get some appreciation of the depth and long history of this sort of problem by considering some examples from philosophy and from an old fable.

In one of his works, *The Republic,* Plato described the Allegory of the Cave. This story describes several people who are chained within a cave such that they can see only the back wall of the cave. They observe shapes moving on the wall and, in trying to understand what is going on, develop explanations about the various shapes and their movements. Eventually, one of the people gets free and, turning around, sees that they have been looking at shadows cast by people and animals that have been passing by the entrance to the cave. He then explains to the others that they have not been describing the real world but only a shadow world. Plato uses this story to argue that the world that we observe is not the real world but is distorted and misrepresented by the limitations of our observations. The real world must be dealt with via thought and reason rather than observation.

The fable describes several blind men who come upon an elephant. They try to understand what sort of beast it is. For example, the man who feels its trunk says that it is something like a snake, the man who feels one of its legs says that it is something like a tree, and so on. Clearly, they are all wrong. Their explanation of the elephant is based on their observations, and those observations are limited and faulty. The real elephant is something beyond their observations.

Thus it may be with energy. We observe what we call light, heat, sound, and motion and think that they are all different. Thus we might say that a light bulb produces light and heat while using up electricity. We might also say that somehow the bulb changes electrical energy into radiant (light) energy and thermal energy. Or we might also say that nothing is really changing, the energy just appears to be in different forms. When it is in the wires, we call it electricity; when it is evident to our eyes, we call it light; when it is detected by our touch, we call it heat. We are much like the blind men and the elephant or the people in the cave.

This issue has taken on general interest because of the work by Albert Einstein (1879–1955) in the early part of this century. He showed mathematically that what we call *mass* is directly related to energy through a simple equation:

$$E = mc^2$$

This suggests more than merely a numerical equality. It suggests that mass and energy are the same thing—All the factor c^2 does is to straighten out the units. So rather than say that in a nuclear reactor some amount of mass is changed into a certain amount of energy, one might say that mass is a form or energy as are light, sound, heat, and so on. Thus the form of energy that we call light can appear as the form of energy that we call mass. Amazingly enough, this has been observed to happen when very high-energy light (called gamma radiation) seems to vanish and mass (an electron and a positron) appears. The reverse process can also happen: Mass can disappear and energy (light) appear. If an electron and a positron get close enough, they go out of existence, and the result is the presence of gamma rays. This is called matter-antimatter annihilation. Lest you think that that is science fiction, this exact process is used in a medical imaging technique called positron emission tomography (PET).

So rather than thinking of light turning into heat or chemical energy turning into some other form of energy, we might use another approach: Simply say that *energy exists.* Unfortunately, we cannot directly observe it. We do detect it by various techniques and, as a result, give it various names. However, it is real and, for some reason that we cannot easily state, cannot be created or destroyed. Perhaps energy represents some level of reality that is beyond our direct observation (as in the allegory and the fable discussed above).

MECHANICAL ENERGY

Energy is one of the most difficult concepts in all of physics. The difficulty dates back to medieval times at least. The question to which energy is one of the answers is "Why do things move and what is it about moving things that makes them different from stationary things?"

This is perhaps not the sort of question that would be popular today, but it was in vogue in the early days of the history of physics. There were several answers to this question. Among them was the concept *vis viva*, the spirit of life. It was thought that movement was associated with life, hence our use of the word "animation" to label "moving" images. The *vis viva* was an ephemeral (without substance) stuff that moving objects had and that stationary objects did not have. As an object gained speed, it gained *vis viva*, and as it slowed down, it lost *vis viva*. The *vis viva* could not be directly observed. Notice the similarity to the caloric model.

During the seventeenth century, the idea of quantifying concepts became popular, and the idea of *vis viva* began to evolve into the concept that we now call **kinetic energy:**

$$KE = \frac{1}{2}mv^2$$

So now one would say that a moving object had kinetic energy and a stationary object had none. As an object speeded up, it gained kinetic energy, and as it slowed, it lost KE.

At this point, there must be a digression to deal with units. As we have seen in the past, it is of utmost importance to be able to use a variety of units in real-life problems.

For example, we have had to use inches, feet, miles, meters, and kilometers when discussing length. The same issue arises in working with energy and heat. There are a variety of units, all in common usage. We will be using the following units:

Btu: A British thermal unit is the amount of heat that will change the temperature of 1 lb of water by 1°F; commonly used in reference to air conditioners, furnaces, stoves, and the like.

cal: A calorie is the amount of heat that will change the temperature of 1 g of water by 1°C.

Cal: A Calorie is equal to 1000 calories; used in reference to food and diet.

J: A joule is a newton-meter; standard unit in the SI system.

kWh: A kilowatt-hour is the standard unit for electric energy.

lb ft: A pound-foot is the standard unit in the USA system.

EXAMPLE	ENERGIES (joules)
Nuclear fuel in the sun	10^{45}
Fossil fuel available on the earth	2×10^{23}
Annual use for the United States	8×10^{19}
Krakatoa	6×10^{18}
Annihilation of 1 kg of matter-antimatter	1×10^{17}
Fission of 1 kg of uranium	8.2×10^{13}
Combustion of 1 gallon of gasoline	1.2×10^{8}
One push-up	4000

As we know, even a person who is sitting still is really moving as the earth moves and therefore has kinetic energy. Imagine a 150-lb person sitting at a desk in school. Since the earth is rotating, he is moving at about 1000 miles per hour if the school is in Quito, Ecuador (a city almost directly on the equator). This gives him about 5×10^{6} lb ft of kinetic energy. He is also moving at about 66,000 miles per hour with the earth around the sun. This gives him an additional 2.2×10^{10} lb ft of kinetic energy. (Contrast this to 3×10^{9} lb ft that is a reasonable amount of energy for a residence to use in one day.) Yet the student would not detect any kinetic energy at all. Kinetic energy cannot be directly observed; it becomes apparent only when it changes.

Kinetic energy changes when an object is speeding up or slowing down. The change in the kinetic energy is given by the following equation:

$$\Delta KE = \frac{1}{2}m\left(v_2^2 - v_1^2\right)$$

Notice that ΔKE will be positive if the object is speeding up and negative if the object is slowing down. Thus as an object speeds up, it is gaining kinetic energy (and this gained energy must come from some other form of energy that is decreasing). As the object slows down, it is losing kinetic energy (and this lost energy must show up as a gain in some other form of energy).

Since we believe that energy cannot be created or destroyed, kinetic energy has to come from somewhere when an object speeds up, and it has to go somewhere when the object slows down. For example, in explaining a collision in a game of pool, one might say that the moving cue ball had *KE*. After it hit one of the other balls, it slowed down, so it had less kinetic energy. The ball that it had struck was now moving, so it had gained some of the kinetic energy that the cue ball had lost. There were some problems with this explanation. Where did the cue ball get its original kinetic energy and what happened to the kinetic energy as both the cue ball and the pool ball eventually slowed to a stop? To answer these as well as other questions, other forms of energy were identified.

One of these other forms of energy is the energy that an object may have as a result of its height. By this I do not mean how tall it is but rather how far up or down it is located. This form of energy is called **potential energy (PE)**. Imagine two bowling balls: a black one held about 6 inches above the floor and a red one held 6 feet above the floor. Upon being released, the red one will certainly be moving faster as it gets to the floor than will the black one. Since the red one had more *KE* as it hit the ground, it seems reasonable to picture it as having more energy at the beginning of its descent. This energy, associated with the height of an object, is called gravitational potential energy and is expressed as follows:

$$PE = mgh$$

Thus, if we were to agree to measure height from sea level, a 150-lb person in Hartford, Connecticut (20 ft above sea level) would have about 3000 lb ft of potential energy. If that same person were to go to Mount Washington (6000 ft above sea level), she would have about 9×10^5 lb ft of potential energy. Yet the person on top of Mount Washington would be no more aware of all of this potential energy than the person in Quito, Ecuador, was aware of all of his kinetic energy. Once again, the amount of energy that an object has is not observable, only the change in the amount of energy is observable.

The change in potential energy (ΔPE) would come into play if the height of the cg of the body were changing:

$$\Delta PE = mg(h_2 - h_1)$$

where *h* is the height, or vertical position, of its cg.

Notice that ΔPE will be positive if the object is rising and negative if the object is falling. Thus as an object's height decreases, the object loses gravitational potential energy (and this lost energy must show up as an increase of some other form of energy). As an object's height increases, its gravitational potential energy increases (and this extra energy must come from some other form of energy that is decreasing).

In a situation in which the only forms of energy are kinetic energy and potential energy, we may write conservation of energy as

$$E_1 = E_2$$
$$KE_1 + PE_1 = KE_2 + PE_2$$
$$0 = KE_2 - KE_1 + PE_2 - PE_1$$
$$0 = \Delta KE + \Delta PE$$

Potential energy and kinetic energy, taken together, are referred to as mechanical energy:

$$ME = PE + KE$$

As an object falls, its height (*h*) is getting smaller, so its potential energy is decreasing, and since it is speeding up, its velocity (*v*) is increasing, and so its kinetic energy is increasing. What happens to this energy as the object hits the ground? Joule suggested that its lost potential energy and kinetic energy show up as heat. Therefore the conservation of energy equation as stated above is not complete.

EXAMPLE 3.3

This is an important consideration in dealing with human locomotion. Consider a person who is walking. As I start to take a step, I push off from the ground. The center of gravity (cg) of my foot goes up, and the cg of my body also goes up. Thus my potential energy is increasing, and since my leg is moving, my kinetic energy also increasing. As I complete a step, the cg of the foot goes back down, and the cg of my body also falls slightly; thus I lose potential energy. The kinetic energy associated with my forward motion probably stays constant, but the KE associated with the vertical motion of my cg decreases, and the kinetic energy of my leg decreases. Thus my body gained mechanical energy as the step began and then lost mechanical energy as the step was completed. As you may already know, the gained mechanical energy probably came from chemical energy that was stored in my body. But what happened to the mechanical energy that was lost as the step was completed? As I complete the step, tendons in my leg are stretching, and there is some compression in tissues within the leg and the foot. The tendons and tissue are to some degree elastic and thus will store energy (this will be discussed later in detail; see page 154) as they are distorted, and then this energy will become available as the distortions lessen. Thus some of the kinetic and potential energy that appears to have been lost will actually be regained as the next step is begun. Experiments have shown that only 65% of the kinetic energy of a moving foot is carried over to the next step; the other 35% is lost to heat and must be supplied by metabolism in the leg muscles. Research investigating the efficiency of women carrying heavy weights either on their backs or on their heads has shown that the loss of kinetic energy from step to step may be reduced from 35% to as little as 15% by those who carry heavy weights on their heads (Zimmer, 1995, p. 28).

If I were to run rather than walk, the loss of mechanical energy would be greater, since my vertical motion is greater. The mechanical energy that I am losing shows up as heat in the joints of my lower body and in the ground.

As we shall see later, bicycling is much more efficient than running. This is because much less loss of kinetic energy is associated with the motion of the legs in riding a bicycle as compared to running. If this "lost" energy could be captured, stored, and then returned to the body, the resulting locomotion would be much more efficient. This is what occurs in jumping on a trampoline or using a pogo stick.

PROBLEM SET 7

7.1.   A 4500-lb car accelerates from rest to 60 mi/h. Determine its change in kinetic energy. (5.45×10^5 lb ft)

7.2.   A 4500-lb car drives up a 3.25-mile hill. The angle of the hill is 5°. Determine the change in potential energy of the car (see figure on p. 140). (6.73×10^6 lb ft)

7.3.   A 70-kg person climbs 15 flights (10 ft each) of stairs. Determine the change in potential energy of the person. (3.14×10^4 J)

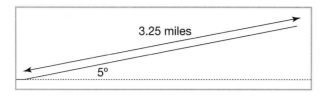

7.4. As an object falls through the air, it is losing potential energy. Usually, the object speeds up as it falls, and so its loss in potential energy is accounted for by the increase in its kinetic energy. However, when things fall through the air, they do not continuously increase in speed. Such an object will reach a maximum speed (called the terminal velocity) and will continue with this speed until it hits the ground. Since the object is continually losing potential energy as it falls, where is this energy going after the object has reached its terminal velocity?

7.5. Why is it reasonable to assume that the mechanical energy of a system before a collision would be greater than the mechanical energy of the system after the collision?

CONSERVATION OF ENERGY

In analyzing a problem, it is necessary to determine the types of energy that are involved. In discussions relating to the human body, we will deal with mechanical energy, chemical energy, thermal energy, and work.

It will be necessary to keep track of the various changes in energy that may play a role in a given problem. One way to do this is to use the following **conservation of energy** equation (*Note*: Review the discussion related to keeping track of the changes in the amount of water in a house):

$$E_{\text{added to the object}} = \Delta PE + \Delta KE + W_{\text{done by the object}} + TE_{\text{leaving the object}} + \Delta U_{\text{within the object}}$$

It is important to note that this equation applies during a specific interval of time between two instants: t_1 and t_2. For example:

a. t_1 might be when a pitcher starts throwing a ball, and t_2 might be when the ball gets to the batter.

b. t_1 might be when a car is at the top of a hill, and t_2 might be when the car is at the bottom of the hill.

c. t_1 might be when a person starts to pull herself up while doing a chin-up, and t_2 would be at the end of the chin-up when she has pulled herself up.

The important idea is that **t_1 is an earlier instant and t_2 is a later instant.** The subscripts 1 and 2 always mean earlier and later, respectively.

$E_{\text{added to the object}}$ represents energy that is added to the object. For example:

a.   In dealing with the human body, this term is usually represented by food that is eaten.

b.   In dealing with an automobile, this term is represented by the amount of fuel that is added to the car at a filling station.

ΔKE is the increase in kinetic energy of the object and would be associated with a situation in which the object ends up going faster than when it started. If the object is slowing down, this term is negative. Notice that we consider only what happens at the endpoints, not what happens between them.

ΔPE is the increase in potential energy of the object and, for our purposes, will be associated with a situation in which the object ended up higher than when it started. If the object ended up lower than when it started, this term would be negative. If the object ended up at the same height as the one from which it started, this term would be zero. Notice that, as with ΔKE, only the endpoints are considered.

$W_{\text{by the object}}$ is the amount of work done by the object on something else. This means that energy is leaving the object and going to something else. Work is the transfer of energy that involves a force and an associated displacement. We shall see later that the force of gravity, that is the basis of the potential energy term, will not be included in considering work. **If work were being done on the object rather than by the object, this term would be negative.**

$TE_{\text{leaving the object}}$ is the amount of thermal energy that leaves the object. Heat/TE is the energy that is transferred from one place to another because of a difference in temperature between the two places. Unless some other form of energy, for example, electric energy in a refrigerator, is playing a role, heat/TE always goes from the higher-temperature place to the lower-temperature place. For the human body, this term may not be negative; it must be positive or zero. The human body certainly may absorb thermal energy from its environment. However, this gained thermal energy will not be available to become kinetic energy, potential energy, and so on.

ΔU is the increase in energy stored within the body. If the object's internal (or stored) energy were being used up rather than increased, then this term would be negative. If we are dealing with the human body, this term will refer to the metabolism of stored fat, sugar, protein, or the like or to elastic (electric) energy associated with stretched tendons, and so on. If we are dealing with an automobile, this term refers to the gasoline that is burned within the engine.

Procedure for using the conservation of energy equation:

1.   Identify the object on which attention will be focused.

2.   Identify the two instants 1 and 2.

3.   List the forms of energy that will play a role in the analysis.

4.   Use the equation.

Although each of the terms in the conservation of energy equation has been defined above, we will now discuss some of them in more detail.

Work

In the restricted vocabulary of physics, the word **work** is used to label the transfer of energy from one object to another when a force and a displacement are involved. For situations in which the force is constant,

$$W = FD\cos(\theta)$$

where W is the work done by the force F on the object, D is the displacement of the object, and θ is the angle between the force vector and the displacement vector. If the work done on an object is determined to be a positive number (θ between 0° and 90° or between 270° and 360°), then the energy of the object is increasing, and some other form of energy is decreasing. If the work done on an object is negative (θ between 90° and 270°), then the energy of the object is decreasing, and some other form of energy must be increasing. If the angle between the force and the displacement is 90°, as in the case of centripetal force, no work will be done (no energy will be transferred).

EXAMPLE 3.4

Consider a pitcher throwing a baseball. The force exerted by the pitcher on the baseball is directed toward the catcher, and since the displacement of the baseball is in the same direction, θ would be 0°. When the ball is caught by the catcher, a very different situation is presented. The catcher pushes back on the baseball to bring it to a stop; the force exerted by the catcher is directed toward the pitcher. The baseball is still moving away from the pitcher. Therefore the force exerted by the catcher and the displacement of the baseball point in opposite directions: $\theta = 180°$. The work done by the pitcher is positive; energy is transferred from the pitcher to the baseball. The work done by the catcher is negative; energy is transferred from the ball to the catcher.

EXAMPLE 3.5

Consider someone who is swinging a weight that is tied to a cord, in a circle around his head. As we have already discussed, the force that he applies to the weight is directed radially inward, that is, centripetal force. The displacement of the weight is directed around the circle. In this case, the angle between the force and the displacement is 90°. Since the cosine of 90° is zero, the work done by the person is zero. This shows that no energy is transferred via a centripetal force.

Considering the units of work (newton-meters), we see that confusion may arise because, as we have already seen, torque is also given as a product of force and distance and so might be measured in newton-meters. There are two ways to avoid this potential confusion.

1. The product of force and distance when applied to torque is written with the distance unit first—for example, meter-newton or foot-pound—and when applied to work/energy/thermal energy is written with the force unit first—for example, pound-foot.

2. The product of newtons times meters when applied to work/energy/thermal energy is renamed the joule.

There is a potential ambiguity that can lead to great confusion in dealing with work and energy. We have defined work as energy that is transferred by means of a force and an associated displacement. However, we have also defined potential energy in terms of the weight of (the force of gravity on) an object. To avoid confusion (and redundancy), **we will limit work done on or by an object to representing forces other than the force of gravity acting on that object.**

EXAMPLE 3.6

A 110-lb hiker is climbing a mountain while carrying a 40-lb backpack. The trail is 10 miles long, and the average angle of elevation is 3°. While she is hiking, she eats an energy bar that is advertised to yield 500 Calories. Discuss the change in energy of her body.

Solution

The object of interest is the hiker's body. Since she is climbing a mountain, her PE is changing. It is reasonable to assume that any changes in her speed while she is climbing are negligible, so ΔKE will be zero. She must be exerting a force on the backpack. If this force exceeded the weight of the pack, it would accelerate upward, and if the force were less than the weight, it would accelerate downward. The only reasonable possibility is that the force equals the weight of the pack and would be directed upward. The energy bar would be represented by energy added to the hiker's body. She would be producing TE as she hiked, and this would have to leave her body, or else her temperature would increase. It is possible that her body will be consuming either more or less energy than that represented by the energy bar. Therefore it is reasonable to assume that her internal energy is changing.

The conservation of energy equation for this problem would be

$$E_{\text{energy bar}} = \Delta PE_{\text{hiker}} + W_{\text{done by hiker on backpack}} + TE_{\text{leaving the hiker}} + \Delta U_{\text{within the hiker}}$$

$$\Delta PE_{\text{hiker}} = (mg)(\Delta h)$$

$$= (110 \text{ lb})(10 \text{ mi})(\sin 3°)\left(\frac{5280 \text{ ft}}{1 \text{ mi}}\right)$$

$$= 304 \text{ k lb ft}$$

$$W_{\text{by hiker on pack}} = (F)(D)(\cos \theta)$$

$$= (40 \text{ lb})(10 \text{ mi})\left(\frac{5280 \text{ ft}}{1 \text{ mi}}\right)(\cos 87°)$$

$$= 111 \text{ k lb ft}$$

$$E_{\text{energy bar}} = (500 \text{ Cal})\left(\frac{1000 \text{ cal}}{1 \text{ Cal}}\right)\left(\frac{0.738 \text{ lb ft}}{0.239 \text{ cal}}\right)$$

$$= 1604 \text{ k lb ft}$$

We shall see later (in our discussion of efficiency) that it would be reasonable to assume that in an activity such as strenuous backpacking, the mechanical energy output (in this case $\Delta PE_{\text{hiker}} + W_{\text{on the pack}}$) is about 10% of the consumed energy (in this case $E_{\text{energy bar}} - \Delta U_{\text{body}}$):

$$\Delta PE_{hiker} + W_{on\ the\ pack} = 0.1(E_{energy\ bar} - \Delta U_{body})$$
$$304\ k\ lb\ ft + 111\ k\ lb\ ft = 0.1(1604\ k\ lb\ ft - \Delta U_{body})$$
$$\Delta U_{body} = -2546\ k\ lb\ ft$$

Since ΔU_{body} comes out to be a negative number, the hiker's body is using up stored chemical energy.

Substituting back into the conservation of energy equation, we may calculate $TE_{leaving\ the\ body}$:

$$TE_{leaving\ the\ body} = 3735\ k\ lb\ ft$$

In summary:

$$\Delta PE_{hiker} = 304\ k\ lb\ ft = 98.4\ kcal$$
$$W_{by\ hiker\ on\ pack} = 111\ k\ lb\ ft = 35.9\ kcal$$
$$\Delta U_{body} = -2546\ k\ lb\ ft = -824.5\ kcal$$
$$E_{energy\ bar} = 1604\ k\ lb\ ft = 519.4\ kcal$$
$$TE_{leaving\ the\ body} = 3.74\ M\ lb\ ft = 1.21\ Mcal$$

Power

The rate at which energy is transferred or at which work is done is called **power**:

$$P = \frac{\Delta E}{\Delta t} \quad or \quad \frac{W}{\Delta t}$$

The SI unit is the joule per second, but this is renamed the watt. The unit of power in the USA system is the pound foot per second. Rather than this cumbersome unit, the horsepower (1 hp = 550 lb ft/s = 746 W) is used. The following chart shows some sample power demands of typical appliances:

APPLIANCE	POWER (W)
Refrigerator/freezer	300
Medium room AC	900
Electric range	8800
Electric dryer	4000
Toaster	1000
Color TV	300
Microwave oven	1500

It is important to note that the "power" does not describe how much work is being done or how much energy is being transferred; the term "power" describes how fast work is being done or how fast energy is being transferred.

A weight lifter may work at the rate of 6 or 7 hp (1 hp = 746 W = 641 Kcal/h) for a couple of seconds, a sprinter at the rate of about 2 hp in a 100-meter race. A trained athlete can keep up the rate of about 0.5 hp for an extended time.

Perhaps the most common occurrence of the term "watt" is for electric devices, such as light bulbs. Consider two bulbs: One is specified as 100 W, and the other is specified as 50 W. One would assume that the former bulb produces twice as much light as the latter. Unfortunately, this may not be the case. What is true is that the first bulb uses electric energy twice as fast as the second bulb. If one were to compare these bulbs to a fluorescent light, the difference is even more striking. A 15-W fluorescent light may produce as much light as a 75-W ordinary bulb. Thus although it is using electric energy only 1/5 as fast as the incandescent bulb, the fluorescent bulb produces the same amount of light.

Note that when you receive a bill for electricity, you are charged for so many kilowatt hours (kWh). Since kW is a unit of power, multiplying by time (hours) gives energy. Thus kWh is a unit of energy, and so you are buying electric energy, not electric power or "electricity," from the utility company. If you use this electric energy in a fluorescent light, you will get more light energy than you would if you used the same amount of electric energy in an incandescent light.

EXAMPLE 3.7

A catcher catches a 115-g baseball that is moving at 80 miles per hour (35.8 m/s) just before it was caught. Assume that the catcher's hand moves backward 35 cm as the ball was stopped.

1. Calculate the magnitude of the force that the catcher's hand exerts on the baseball.
2. Calculate the rate (in hp) at which the catcher is doing work.

Solution

1. The object will be the baseball.
2. The two instants will be when the ball gets to the catcher (t_1) and when the ball has stopped (t_2).
3. The forms of energy involved will be kinetic and work.
4. The basic equation will take the form

$$0 = \Delta KE + W_{\text{done by the baseball}}$$

or

$$0 = \Delta KE - W_{\text{done by the baseball}}$$

Let's use the second form. We need the work done on the baseball by the catcher, so we must consider the force exerted by the catcher on the baseball. The magnitude of the force is unknown (we can call it F_c). However, we know that it is directed toward the pitcher, while the ball is moving away from the pitcher. Therefore the angle between the force and the displacement of the ball will be 180°.

5. Substituting the information into the basic equation, we have;

$$0 = \frac{1}{2}(115 \text{ g})\left(0^2 - \left(80 \frac{\text{miles}}{\text{hr}}\right)^2\right) - ((F_c)(35 \text{ cm})(\cos 180°))$$

6. After dealing with the issue of units, we determine that

$$F_c = 210 \text{ N}$$

To calculate the rate at which the catcher is doing work on the baseball (the output power of the catcher), we use the formula

$$P = \frac{W}{\Delta t}$$

W is the amount of work done by the catcher on the ball. This may be calculated to be −3.5 J. We may ignore the minus sign, as we are interested only in the amount of work done by the catcher. Now we need Δ*t*, the amount of time it takes for the catcher to bring the baseball to a stop. Since we know the initial speed (35.8 m/s) and the final speed (0) of the baseball, we might try to use acceleration. The magnitude of the acceleration could be determined from Newton's Second Law:

$$\vec{F}_{catcher\ on\ ball} = m_{ball}\vec{a}_{ball}$$

From this equation we see that the direction of the acceleration of the ball must be the same as the direction of the force exerted by the catcher (and this is directed toward the pitcher). The velocity of the ball is toward the catcher, and we shall pick this to be the positive direction. Thus the acceleration of the ball will be negative:

$$F_{by\ catcher} = m_{baseball}a_{baseball}$$

$$-210\ N = 0.115\ kg\ a_{baseball}$$

$$a_{baseball} = -1826\ m/s^2$$

$$a = \frac{\Delta v}{\Delta t}$$

$$v_1 = 35.8\ m/s \qquad and \qquad v_2 = 0$$

$$-1826 = \frac{0 - 35.8}{\Delta t}$$

$$\Delta t = 0.02\ s$$

$$P = \frac{73.5\ J}{0.02\ s}$$

$$P = 3.68\ kW = 5\ hp$$

Comparing this power output to the examples mentioned earlier, we see that it is a very large value. A catcher is able to work at such a high rate only because the activity continues for such a short time (0.02 s). No person could work at this rate for an extended time.

INTERNAL ENERGY

The change in internal energy (ΔU) of the object would play a role if there were either an increase or a decrease in energy that is stored within it. Stored energy can take many forms. Some of these are as follows:

a. Thermal energy (sometimes called heat). This is really kinetic energy at the atomic or molecular level (page 136). A change in thermal energy is indicated by a change in temperature (page 175) or a change in physical state of the object (page 176).

b. Chemical energy. This is the energy associated with the electrical interactions of the atoms and/or simple molecules that make up the more complex molecules within the object. A change in chemical energy is indicated by the appearance or disappearance of such complex molecules as due to chemical reactions (page 150).

   c. Elastic energy. This is the energy associated with the electrical interactions of the atoms and molecules that determine the geometry (e.g., shape, length, surface area) of the object. A change in elastic energy is indicated by a change in the geometry of the object (page 154).

   d. Nuclear energy. This is energy associated with the interactions of the protons and neutrons that make up the nuclei within the object. A change of nuclear energy is indicated by a change in the elements that make up the object (page 163).

A general expression for the change in internal energy of an object is

$$\Delta U_{general} = \Delta TE_{\text{within the object}} + \Delta E_{\text{elastic}} + \Delta E_{\text{chemical}} + \Delta E_{\text{nuclear}}$$

In dealing with the human body, while some energy is stored in stretched soft tissue, virtually all of the internal energy is in the form of chemical energy. Therefore:

$$\Delta U_{\text{body}} = \Delta E_{\text{chemical}} + \Delta E_{\text{elastic}}$$

is usually

$$\Delta U_{\text{body}} = \Delta E_{\text{chemical}}$$

CHEMICAL ENERGY

Chemical energy is the form of energy that becomes evident when a change in the chemical composition of the material occurs. The most common application of this quantity occurs when food or fuel is used to provide energy. In these cases, the material (food or fuel) is chemically combined with oxygen (*oxidation*). If the oxidation process is rapid, it is called *combustion*. Oxidation is an exothermic chemical reaction, which means that as a result of the reaction, thermal energy is produced. We will discuss later how thermal energy can be "produced."

    Chemical energy is important to us because it is the basis of food. Plants absorb radiant energy from the sun. This energy is used (during photosynthesis) when the plants combine simple molecules such as H_2O and CO_2 into complex molecules such as carbohydrates. The energy is stored in the resulting molecules. We may eat these plants, and the molecular structures within them are changed (during digestion) into other complex molecules that may be used by the body. When these complex molecules are broken down (during metabolism) into H_2O and CO_2, energy is released. Thus plants (or, better, the sun) are our basic source of energy.

    Cellulose is a carbohydrate that represents more than 50% of the total organic carbon in the world. Unfortunately, the human body (specifically the small intestine) does not contain the enzymes that are necessary to break cellulose down into usable molecular structures. We rely on other animals (cows, sheep, etc.) to convert cellulose into a form from which we can derive energy. This energy that we take in from animal products makes up only 10–15% of the energy that the animals derived from the plant material. Therefore this process is only 10–15% efficient. Even though we cannot digest cellulose, it is a necessary part of the human diet; it is what is called "fiber."

A person with a healthy digestive system will absorb almost all of the digestible food that is eaten. Specifically, 99% of the carbohydrate, 92% of the fat, and 93% of the protein delivered to the small intestine are normally absorbed (Lenihan, 1975, p. 161). Such a system would be very desirable in situations in which food is not readily available: When someone finds some food, virtually all of it (well over 90%) is absorbed by the body and then is available to supply needed energy. If a person lives in a situation in which food is readily available and eats regularly, over 90% is still absorbed and stored within the body. It should not be surprising that being overweight is one of the recurrent concerns of people in developed countries and is thought to be the basic cause of many serious medical problems, such as hypertension. From the point of view of food as a supply of required energy, it seems clear that we eat too much. The process of eating has gone beyond satisfying energy needs and now is related to other, perhaps emotional, needs.

The cells that make up the body require energy to function. For example, chemical energy (in the form of sugar) is absorbed by a cell and is then (by action of the mitochondria) converted to chemical energy, represented by ATP. The tissue that is composed of these cells is thus dependent on taking in energy. Therefore energy is required by muscle tissue whenever it contracts, by neurons whenever they fire, and by many chemical processes. So there is a demand for energy by the body even when there is no exertion, as during sleep. The heart is still beating, and since it is a muscle undergoing successive contractions, it consumes energy. The liver, carrying out continuous chemical processes, also consumes energy continuously. The brain, parts of which are always active, is also a continuing consumer of energy. This minimum rate of energy consumption is called the **basal metabolic rate (BMR)**. A typical value of the BMR is about 107 watts (91 kcal/hour) and decreases by about 10% for each C° drop in temperature. (*Note*: Since the BMR represents the rate at which chemical reactions are proceeding, the decrease in BMR with lowered temperature is consistent with the statement that, in general, chemical reactions take place more slowly at lower temperatures.)

This explains why warm-blooded animals can be more active than cold-blooded animals in cold weather. The internal temperature of warm-blooded animals stays constant even if the ambient temperature falls. Therefore the rate at which they use stored chemical energy also stays constant. For cold-blooded animals, the rate at which stored energy is used decreases as the ambient temperature falls. They have less energy available and become torpid. Thus animals whose internal temperature may stay high, even in cold surroundings, have a distinct advantage over other animals. They can move faster for longer times. Some animals that are usually considered to be cold-blooded can actually have an internal temperature that is higher than that of their surroundings. Examples include tuna fish and sharks (Carey, 1973). These animals can swim faster for longer periods of time than other sea animals on which they prey.

Measurement of the rate of energy consumption

In 1784, Lavoisier suggested that the process by which we extract energy from food is oxidation. The most important chemical process that liberates chemical energy within our bodies is the oxidation (or combustion) of a particular sugar: glucose. This reaction is represented by the following equation:

$$C_6H_{12}O_6 + 6O_2 \longrightarrow 6H_2O + 6CO_2 + \text{energy}$$

Specifically, 1 mole of glucose (180 g) combines with 6 moles of oxygen (192 g) to produce 6 moles each of water (108 g) and carbon dioxide (264 g) while releasing 686 kcal of chemical energy. This energy typically shows up as thermal energy and mechanical energy.

In principle, we could determine how much energy the body is using by measuring the amount of glucose that is consumed. This is not possible because glucose is being consumed by cells all over the body, and therefore the process cannot be localized and monitored. However, we can measure the amount of oxygen that the body consumes. It is relatively easy to measure the amount of oxygen that is inhaled and the amount that is exhaled. Subtraction yields the amount of oxygen consumed, called the oxygen uptake.

This technique has been used to measure energy consumption for sports training purposes, for medical diagnosis, and for research in animal physiology with such diverse animals as caterpillars and elephants. The figure below shows an example of this type of research (Discovery Magazine, July 1995, p. 18). The elephant is wearing a mask that controls both the inhaled and exhaled air. There is a monitor on the vehicle that measures the rate at which the elephant is inhaling oxygen and the rate at which oxygen is being exhaled. Thus the net rate of oxygen consumption may be determined. This, in combination with the material discussed previously, then leads to a determination of the rate at which the elephant is consuming energy.

Using the equation, we can relate the oxygen uptake in liters to the amount of chemical energy released. We know that 686 kcal of energy are released for every 6 moles of oxygen consumed; therefore (assuming that 1 mole of oxygen occupies 22.4 liters of volume at standard conditions of temperature and pressure) we have:

$$\frac{686 \text{ kcal}}{(6 \text{ mol} \times 22.4 \text{ L/mol})} = 5.1 \frac{\text{kcal}}{\text{L}}$$

Thus for each liter of oxygen uptake, 5.1 kcal of chemical energy are consumed by the body and are therefore available for conversion into some other form(s), such as thermal or mechanical energy.

Once it becomes possible to measure the rate at which the body is consuming energy, one may determine the efficiency of the body in doing a known amount of work. In this type of research, the amount of energy that is consumed is measured, the amount of work done is measured, and the ratio of these is calculated. This ratio,

$$\frac{\text{amount of work performed}}{\text{amount of energy consumed}}$$

is the efficiency of the body in carrying out this particular form of work. (Efficiency will be discussed in detail in a later section.) This type of measurement is used in training programs to help an athlete to be able to produce more mechanical energy during an athletic competition or perhaps to convert chemical energy into mechanical energy more rapidly.

This sort of measurement has produced some rather startling information. For example:

1. It has been experimentally determined that the change in the metabolic rate of humans, ponies, and dogs is proportional to the amount of weight that they carry (expressed as a percentage of body weight). This means that if a person were to carry a weight equal to 50% of his or her body weight, the person's metabolic rate would increase by 50%. However, when the rhinoceros beetle is examined in a similar experiment, it is found that when the weight carried is as much as 30 times (3000%) the body weight, the metabolic rate increases by only 400% (Zimmer, 1996).

2. In investigations of the relative energy costs of self-propelled locomotion of various animals, it has been determined that if a mouse and a sparrow (which have approximately equal masses) are moving so that they have equal rates of metabolism, the sparrow will be moving 10 times as fast as the mouse (Tucker, 1969).

In the human body, internal energy is stored as chemical energy in the forms of carbohydrates (sugar), fats, and protein. Fats constitute a much more efficient way of storing energy than do carbohydrates. This is because the oxidation of fat produces about twice as much energy per gram as does the oxidation of carbohydrates or proteins. Storing energy in the form of carbohydrates presents another problem. To be oxidized, each gram of carbohydrate requires 4 to 5 grams of water. So the amount of weight required per unit of energy produced increases dramatically. However, some energy storage in the form of carbohydrates is important. Glycogen is stored as granules in skeletal muscle fibers and in liver cells. It is converted to energy during intense activity, and liver glycogen is used to maintain blood glucose levels.

We will represent the amount of chemical energy liberated in an oxidation reaction by

$$\Delta E_{\text{chemical}} = (\text{quantity of the material})(\text{heat of combustion})$$

where the heat of combustion (H_c) for a particular material may be found in a table (see below) and the quantity of material will be the number of grams, pounds, gallons, or the like, as required.

In general, the oxidation of carbohydrates and proteins produces 4 kcal/g, of lipids produces 9 kcal/g and the oxidation of alcohol produces 7 kcal/g (Davidovits, 1975, p. 134). More specifically, we have the following values for H_c:

FOOD OR FUEL	EQUIVALENT ENERGY H_c
Carrots	45 kcal/cup (Davidovits, 1975, p. 135)
Chocolate chip cookie (1)	35 kcal
Coal	3632 kcal/lb
Doughnut (1)	135 kcal (Davidovits, 1975, p. 135)
Egg (1)	75 kcal (Davidovits, 1975, p. 135)
Fat	9.3 kcal/g (Urone, 1986, p. 74)
Food oil	3500 kcal/lb
Gasoline	12×10^7 J/gal = 46 kJ/g
Ice cream	2.22 kcal/gram
Lean beef	1000 kcal/lb
Sugar	1600 kcal/lb
TNT (1 megaton)	4×10^{12} kJ
Whole milk	660 kcal/quart (Davidovits, 1975, p. 135)

EXAMPLE 3.8

If a person were to metabolize 7 grams of fat (the fat content of four chocolate chip cookies), the amount of energy produced would be given by

$$\Delta E_{fat} = m_{fat} H_{c,fat}$$

$$\Delta E_{fat} = (7g)\left(9.3 \, \frac{kcal}{g}\right)$$

$$\Delta E_{fat} = 65.1 \text{ kcal}$$

To understand how energy can be produced from a chemical reaction, we must consider what happens at the molecular level. We know that a molecule is made up of atoms that in turn are composed of positively charged nuclei surrounded by clouds of negatively charged electrons. We can thus imagine that the molecules are surrounded by such negatively charged clouds. Even though the molecules themselves may be electrically neutral, when they approach each other, their negatively charged regions interact first. Thus they inherently repel each other. We can represent this phenomenon by imagining that each molecule has, sticking out from its center, many springy coils. As the molecules approach each other, these springs interact, become compressed and tend to push back, forcing the molecules apart. However, if there were a mechanism that somehow clamped the molecules together, thereby compressing the springs when the molecules were close, energy would be stored. This would become evident when the mechanism was removed. The springs would expand, forcing the molecules apart. The molecules would move apart, thus displaying kinetic energy. If the moving molecules were then slowed down, the kinetic energy would appear as thermal energy. Thus the energy that had been used to get the molecules close enough together for the mechanism to work would now show up as thermal energy. Heat would appear to have been produced, but no energy would really have been created. Energy had been stored in the system when the molecules were forced close together, and this energy later reappeared. Referring back to the chemical reaction, we can imagine that the molecules are CO_2 and H_2O. They are brought together by the process of photosynthesis that uses radiant energy from the sun. They link up, forming glucose ($C_6H_{12}O_6$) and oxygen (O_2). When we eat the plant material containing glucose, enzymes in our digestive system break apart the links, and the $C_6H_{12}O_6$ combines with O_2 to form CO_2 and H_2O and, as with the "springy molecules" described above, produces energy. This step is called oxidation and is exothermic. Note however that **energy was not created. The energy that was used to bring the molecules together was stored in the new structure and was then released when this structure was disassembled.**

The rate at which the human body consumes energy can be expressed in many ways. The method adopted by the American Heart Association is to use as a basis the MET. This is a unit of sitting, resting oxygen uptake (3.5 mL O_2 per kg of body weight per minute [mL kg^{-1} min^{-1}]. One may calculate the **rate of energy consumption** by using (American Heart Association, 1990)

$$P_{consumed} = (\text{number of METs})(0.0175)(\text{body weight in kg}) \frac{\text{kcal}}{\text{min}}$$

(*Note*: The quantity 0.0175 is required to make the units come out correctly.)

ACTIVITY	POWER$_{consumed}$ (MET)	POWER$_{consumed}$ (kcal/min/kg)
Walking (2 mph)	2.5	0.044
Cycling (leisurely)	3.5	0.061
Swimming (slowly)	4.5	0.079
Walking (4 mph)	4.5	0.079
Cycling (moderately)	5.7	0.1
Swimming (fast)	7.0	0.12
Jogging (10 minute mile)	10.2	0.18
Rope skipping	12.0	0.21
Squash	12.1	0.21

EXAMPLE 3.9

Consider a 105-lb person who is skipping rope for 1 hour. Assume that all of the energy that her body uses comes from the combustion of glucose. How much glucose is consumed?

Solution

According to the chart, rope skipping requires 12.0 METs. Using the formula above the chart, we have

$$P_{consumed} = (\text{METs})(0.0175)(\text{body weight in kg}) \text{ kcal/min}$$
$$P_{consumed} = (12.0)(0.0175)(49.9) \text{ kcal/min}$$
$$= 10.48 \text{ kcal/min}$$
$$E_{consumed} = P_{consumed} \, \Delta t$$
$$= (10.48 \text{ kcal/min})(60 \text{ min})$$
$$= 629 \text{ kcal}$$
$$E_{consumed} = m \, H_c$$
$$629 \text{ kcal} = m(1600 \text{ kcal/lb})$$

Thus about 0.4 lb of glucose is consumed.

EXAMPLE 3.10

While doing a sit-up, a 75-kg person raises the center of mass of her upper body by 25 cm. While she does so, her abdominal muscles contract (shorten) by 2 cm.

1. Determine the change in her mechanical energy.
2. Did her mechanical energy increase or decrease? How do you know?
3. If you determined that her mechanical energy increased, where did the extra energy come from? If you determined that her mechanical energy decreased, where did the lost energy go?
4. How much force was exerted by her abdominal muscles as she did the sit-up?
5. Express this force as a multiple of her weight.
6. How many sit-ups would she have to do to burn off the energy represented by a 1/4 pint of chocolate ice cream (weight of 1 pint of ice cream is 0.8 lb)?

Solution

1. Her change in mechanical energy would be the change in the potential energy of the upper part (trunk, head, neck, and arms) of her body (which constitutes 66.1% of her body weight):

$$\Delta PE = mg \, \Delta h$$
$$= (0.661)(75)(9.8)(0.25) = 121.5 \text{ J}$$

2. Her mechanical energy increased because, since the center of gravity of her upper body was rising, PE increased.
3. The increase in mechanical energy came from a decrease in stored chemical energy.
4. The change in potential energy was caused by work done by her abdominal muscles. Her muscles did work on the upper part of her body. We can use the conservation of energy equation:

$$0 = \Delta PE_{\text{upper body}} - W_{\text{on upper body}}$$
$$0 = \Delta PE_{\text{upper body}} - W_{\text{by abdominal muscles}}$$

The work done by her muscles is $W = FD \cos \theta$. Since muscles always pull and these muscles are contracting, $\theta = 0°$:

$$0 = 121.5 - F(0.02) \cos (0)$$
$$F = 6075 \text{ N}$$

5. $\dfrac{F_{\text{muscles}}}{W_{\text{body}}} = \dfrac{6075}{(75)(9.8)} = 8.3$

6. Her body gains chemical energy by metabolizing the ice cream. This amount of energy is given by

$$\Delta E_{\text{chemical}} = (\text{quantity of the material})(\text{heat of combustion})$$
$$\Delta E_{\text{chemical}} = \left(\frac{1}{4} \text{ pt}\right)\left(\frac{0.8 \text{ lb}}{1 \text{ pt}}\right)\left(\frac{1000 \text{ g}}{2.21 \text{ lb}}\right)\left(\frac{2.22 \text{ kcal}}{1 \text{ g}}\right)$$
$$= 201 \text{ kcal} = 841 \text{ kJ}$$

Since she uses 121.1 J for each sit-up, she must do almost 7000 sit-ups. (*Note*: This number is not realistic because we have not taken into account the thermal energy that her body produces as she does the sit-ups. We will deal with this in the section on efficiency.)

ELASTIC ENERGY

Knowledge of the elastic properties of human tissue is vital to an understanding of the functioning of the body. If you were to jump from a chair down to the ground and land stiff-legged, you would feel quite a shock as your feet hit the ground. There is relatively little give in rigid legs, so your body would stop in a very short time and therefore undergo a large Δv in a short Δt. This implies a large acceleration, and the body therefore experiences a correspondingly large force. You might actually suffer broken bones by landing in this manner. If, on the other hand, you were to land with your legs relaxed so that they could flex when your feet hit the ground, you would feel much less shock, and there would be less chance of bones breaking.

As your legs flex, the quadriceps muscles and tendons stretch. The quadriceps act such as shock absorbers. They take time to stretch. Therefore the acceleration is less, and so less force acts on your body. Also, energy is required to make the quadriceps stretch, and this energy is eventually dissipated as thermal energy.

When you stretch a spring, you do work, transferring energy to the spring. As a result, its internal energy increases. If you allow the spring to contract and it returns to its original length, the energy that was stored must be accounted for (conservation of energy). In an ideal situation, all of the stored energy would show up as mechanical energy (probably kinetic energy) as the spring shortened. This means that none of the stored energy would be lost as thermal energy. In a real spring, some of the stored energy will always show up as thermal energy. Biological tissue, particularly muscles and tendons, behave in the same way: They store energy when stretched by an external force. Some of this energy may show up as mechanical energy when the external force is removed and they return to their original shape. Unfortunately, only a very small fraction of the body's mechanical energy that is absorbed by bone and soft tissue is later converted back to mechanical energy. Most of the absorbed energy shows up in thermal form. (We will return to this topic later.) Therefore the elastic properties of the soft tissue that cushions the vertebrae and joints such as the ankle, knee, and hip, as well as the muscles and tendons, play an important role in absorbing mechanical energy. This helps to reduce the effects on these bones each time we take a step, run, or jump.

You have probably felt a sort of bounce when you run or jump while wearing high-quality sports shoes. This is caused by the ability of the material in the shoes' soles to store some of the mechanical energy that the body loses with each step and then to convert that energy back to mechanical energy for the next step. An extreme example of this is evident when a person jumps repeatedly on a trampoline. The springs that support the trampoline surface absorb the body's mechanical energy on impact and then give a large fraction of this energy back to the body on the next jump. So you can go higher and higher, not only because your body is using up more internal energy but also because, during each jump, the trampoline is giving some of your previous mechanical energy back to you.

Biological materials vary greatly in their elastic properties. There is a protein, appropriately called *elastin* (Fung, 1997, pp. 243–245), that is very elastic and can absorb, store, and then give back energy. Elastin is a component of some soft tissue (e.g., skin and blood vessels) within your body. But there is not enough of it to make a large contribution to the mechanical energy requirements of your body. The elastin in your skin gives it the ability to remain smooth. If a young person's skin is stretched or otherwise distorted and then released, it quickly returns to its natural state. Unfortunately, the ability of the human body to produce elastin decreases after puberty (Fung, 1997, p. 244). So as we age, skin loses its elasticity, and permanent wrinkles appear. The largest concentration of elastin in the body is in the blood vessels, for example, in the aortic arch. This structure is located at the top of the heart. Blood that has been oxygenated passes from the heart to the rest of the body through this arch. Three elastic arteries originate along the arch. They carry blood to the head, neck, shoulders, and arms. There is a high concentration of elastin in the arch and in these arteries. When the heart contracts, chemical energy in the muscle tissue is converted to mechanical energy, represented by the moving blood. As this pulse of moving blood passes through the arch and into the arteries, they expand in response to the increased blood pressure. The tissue stretches, absorbing kinetic energy from the blood and storing it as elastic energy. As the pulse passes, the tissue contracts, squeezing the blood. This squeezing does work on the blood, increasing its kinetic energy. Thus some of the elastic energy is returned as mechanical energy. It is the high concentration of elastin in this tissue that makes this process efficient.

Another elastic protein, resilin, is found in the wing and leg structures of some insects. This material has the amazing property that it can return as much as 80% of its elastic energy as mechanical energy. This helps to explain how a creature as small as an insect can fly. Its wings are continually changing velocity: speeding up, slowing down, changing direction, then speeding up again, and so on. Their kinetic energy is therefore also changing: increasing and decreasing. It might be expected that when the kinetic energy of the wings increases, an equivalent amount of metabolic (chemical) energy must be expended, and when the kinetic energy decreases, an equivalent amount of thermal energy would appear. This is not possible because the required loss of chemical energy would exceed the amount that such a small creature could reasonably store, and the associated rate of thermal energy production would exceed the ability of such a small creature to dissipate.

Resilin in the wing structures provides part of the solution. As the wings slow down, approximately 80% of the lost kinetic energy is converted into elastic energy within the resilin tissue. Therefore less thermal energy is produced. At the beginning of the next flap of the wings, most of this elastic energy is converted back into kinetic energy. Thus the wings pick up some kinetic energy from stored elastic energy rather than using up the equivalent amount of metabolic energy.

To begin a quantitative analysis of the forces and energies associated with the deformation of materials, we must first discuss some of the basic physical principles. Since the shape of an object, such as a bone or a tendon, is determined by the electrostatic forces between the molecules that compose it and since these forces are limited in size, it is understandable that too much distortion can cause the object to fail, that is, to break or tear. The maximum distortion that an object can withstand without failing is called the **breaking strain.** (There are other terms used for this property, such as ultimate elongation

(Park, 1984, p. 160).) Bone is 60–70% calcium phosphate, and calcium carbonate by weight. These materials, when under compression, have a large breaking strain. This is why bone is able to support your body by pushing up against the force of gravity. The other major component of bone is a protein, *collagen*. Collagen makes up 60% of the volume of the bone. It has a high breaking strain against tension. This gives bone some ability to withstand twisting and bending. Thus, bone is a two-phase, or composite, material. It is composed of two very different materials; one (the compressive phase) is able to withstand high compression without failing, and the other (the tensile phase) is able to withstand tension without failing. There are many examples of two-phase materials in the world. Reinforced concrete is composed of concrete, which provides strength against compression, and steel rods, which provide strength against tension. The graphite epoxy materials used in bicycle frames and tennis rackets are similar. Even though bone contains a tensile phase, its main use in the body is to push, that is, to withstand compression. When bone breaks, it is usually because of tensile failure, not compressive failure. On the other hand, soft tissue such as muscle and tendon can withstand tension but have almost no ability to withstand compression. They can pull, but they cannot push. This is why the major skeletal muscles in your body are arranged in opposing sets: one to provide abduction and the other adduction, or one set for flexion and the other for extension. If muscles could push, that is, withstand compression, we would not need the opposing sets.

To summarize, bones withstand compression (they push) and soft tissue withstands tension (it pulls).

If an object or a piece of material returns to its original form after the force is removed, such as a rubber band or human skin, the material is said to be elastic. If it does not return to its original form, it is said to have been plastically deformed. Modeling clay is a good example of a plastic material. Note that, oddly enough, most pieces of what we call "plastic" are not plastic in their response to an applied force.

The degree to which a material will respond to an applied force in an elastic or plastic manner may depend on the amount of time during which the force is applied. For example, if a child's femur is subject to a force during a short time interval, as when hitting the ground after jumping, it will probably react elastically. This means that it will momentarily shorten or bend on impact and then return to its original length when the force diminishes. On the other hand, a child who is bowlegged may be helped by a properly designed brace that applies a steady force to the neck of the femur, causing it to bend. If the brace is worn for a sufficiently long time, the femur's shape will be permanently changed, remaining properly bent even after the brace is removed.

The terms used to describe deformations are **stress** and **strain**. **Stress** is the ratio of the applied force to the cross-sectional area over which the force is applied. It is usually measured in newtons per square meter (N/m^2 or Pa), dynes per square centimeter, or pounds per square inch:

$$stress \equiv \left(\frac{F}{A} \right)$$

Strain is the fractional deformation, for example, the change in length divided by the original length:

$$\text{strain} \equiv \left(\frac{\Delta L}{L_o}\right)$$

Since the strain is a ratio of two such quantities (those having the same units), the dimensions cancel, and the ratio is dimensionless. It is often expressed as a percentage.

Each kind of material is characterized by the ratio of stress to strain. For example, in dealing with tension or compression, Young's modulus is used. It is defined by

$$Y = \frac{\text{stress}}{\text{strain}} = \frac{\left(\dfrac{F}{A}\right)}{\left(\dfrac{\Delta L}{L_o}\right)}$$

where F is the applied force, A is the cross-sectional area, ΔL is the amount of distortion (stretch or compression), and L_o is the original length. Thus we see that, for a given amount of stress (force), a greater Young's modulus will yield a smaller amount of strain (a smaller change in length).

The graph below shows how the strain is related to the stress in a typical material. The region between the origin and point A is characterized by a large amount of strain (distortion) resulting from a small amount of stress. In the case of soft biological tissue, such as tendon or ligament, this is called the *toe region*. As the material is stressed through this region, microscopic internal "crimps" are stretched out (Nigg and Herzog, 1995, p. 116). The material is reacting to the applied stress by small, localized distortions rather than as

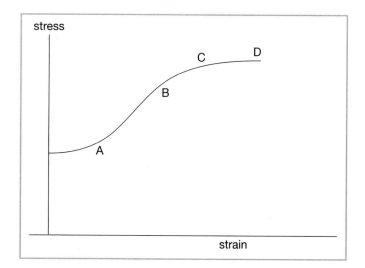

a whole piece. As the stress passes point *A*, the relative strain decreases. The material is reacting as a whole. The region between *A* and *B* is called the linear region. As the stress exceeds *B*, microscopic internal failures start to occur within the material, and the relative strain increases. Beyond *C*, larger-scale failure takes place, and at *D*, the material tears. The region *CD* is called the *plastic region*. A material distorted into this region will not return to its original shape when the stress is removed. The strain at *D* is the breaking strain, which was defined above.

From the point of view of mathematics, the slope of the graph would be the ratio

$$\text{slope} = \frac{\Delta\text{stress}}{\Delta\text{strain}}$$

It should be clear that this is Young's modulus.

Consider the region *AB*. The graph is linear; the slope is constant. Thus in the region *AB*, *Y* is constant. Solving the Young's modulus equation for *F*, we have

$$F = \frac{YA}{L_0}\Delta L$$

When an object is stretched, its cross-sectional area, *A,* will decrease. However, for most situations, the change in area is so small that it may be ignored in the analysis of the distortion. In this case, recalling that L_0 is the original length of the object, the quantity YA/L_0 will remain constant. It is called the spring constant and is designated by the letter *k*. The equation for the force may then be written as:

$$F = k\,\Delta L$$

This equation represents the amount of distortion (ΔL) as being directly, or linearly, proportional to the amount of force (*F*) that is applied to the object. This means that if we double the amount of force, the amount of distortion will also double. This condition defines what is called the *elastic region* or *Hooke's law region*.

As one progresses from the head down to the feet, the amount of weight compressing the next layer of bone increases. This steadily increasing force should result in a steadily increasing stress. Since the value of *Y* is the same for the various bones, we would expect much more distortion in lower bones than in upper bones. However, as you know, in general, the cross-sectional area of our bones increases from head to foot, and therefore the stress, the ratio of force to area, does not increase substantially. Those lower bones whose cross-sectional areas are not large, such as those in the ankles or feet, are subject to much larger stresses than other bones, such as the femur, and are more likely to fail under compression, as with compression fractures in a runner's feet. Hands of boxers and weight lifters are also subject to breakage for the same reason.

The table lists relevant information for some common materials. Notice that bone can withstand almost five times more stress against compression than against twisting. Also notice that steel has a much larger Young's modulus than does bone. Therefore if the stresses are equal, a steel implant prosthesis will distort much less than the bone that it replaced. This is an important consideration in the design of such implants.

MATERIAL	YOUNG'S MODULUS $(10^9$ Pa$)$	BREAKING STRESS $(10^6$ Pa$)$
Steel	200	450
Aluminum	69	62
Bone (femur)	20(Benedek and Villars, 1973), 14 (Davidovits, 1975; Hobbie, 1988)	100 (Davidovits, 1975), 180 (Hobbie, 1988) (compression)
		121 (Benedek and Villars, 1973), 83 (Davidovits, 1975; Hobbie, 1988) (tension) 27.5 (Davidovits, 1975) (twist) 208 (Benedek and Villars, 1973) (bending)
Vertebral column		4.6 (compression, males $<$ 60 years old) (Duncan, 1990) 3 (compression, males $>$ 60 years old) (Duncan, 1990)
Tendon	1.5 (Nigg and Herzog, 1995, p. 144)	68.9 (Davidovits, 1975) (tension)
Muscle		0.55 (Davidovits, 1975) (tension)
Elastin	0.0004 (Fung, 1997, p. 244)	
Resilin	0.0018 (Fung, 1997, p. 244)	

Note: 1 Pa = 1 N/m^2.

We can now deal with elastic internal energy. The change in internal energy associated with an elastic deformation in the Hooke's law region is given by

$$\Delta E_{\text{elastic}} = 0.5k(\Delta L)^2$$

We may substitute for the spring constant from above to get

$$\Delta E_{\text{elastic}} = \frac{YA(\Delta L)^2}{2L_o}$$

Let us denote the breaking stress by S_b. The maximum amount of compression is given by:

$$\Delta L_{\text{max}} = \frac{S_b L_o}{Y}$$

The amount of energy associated with this amount of compression is:

$$\Delta E_{\text{elastic,max}} = \frac{A S_b^2 L_o}{2Y}$$

The product AL_o is the volume of the material. Dividing through by this quantity will give an equation for the material to change mechanical energy into elastic energy per unit volume u_{max} in J/m^3:

$$u_{max} = \frac{S_b^2}{2Y}$$

EXAMPLE 3.11

Compare the relative abilities of steel, *femur*, and tendon to absorb mechanical energy.

Solution

MATERIAL	$Y\,(10^9\,\text{Pa})$	$S_b\,(10^6\,\text{Pa})$	$u_{max}\,(10^5\,J/m^3)$
Steel	200	450	5.1
Femur (compression)	17	150	6.6
Tendon	1.5	70	16.3

This shows that tendon can absorb twice as much energy (per unit volume) as bone. Unfortunately, this amount of elastic energy will probably not be available for conversion back into mechanical energy. Most of it will show up as thermal energy within the material. This will be discussed in more detail later.

EXAMPLE 3.12

Determine the maximum height from which a 150-lb person may fall and land stiff-legged on both feet without breaking his tibias or femurs. We can assume that the lost potential energy will show up as an increase in $\Delta E_{elastic}$.

Solution

$$E_{\text{added to the object}} = \Delta PE + \Delta KE + W_{\text{done by the object}} + TE_{\text{leaving the object}} + \Delta U_{\text{within the object}}$$

$$0 = \Delta PE + \Delta U_{body}$$

$$\Delta U_{body} = \Delta E_{elastic}$$

$$|\Delta PE_{max}| = |\Delta E_{elastic,\,max}|$$

$$mg\Delta h_{max} = \frac{AS_b^2 L_o}{2Y}$$

We see from the table that a reasonable value for the breaking stress of human bone (femur) under compression is $100 \times 10^6\,N/m^2$ and that $Y = 20 \times 10^9\,N/m^2$. Assuming that the femur and tibia have a combined length 0.76 m and average cross-sectional area of 6 cm^2, we find that the maximum height is about 34 cm (17 cm if the person lands on one foot).

Of course, a person may fall from a greater height than this and land without breaking both legs, but that person will probably flex his or her knees and thus absorb some of the energy by stretching tendons and muscles rather than by compressing the bones in the legs.

We will now consider the conversion of stored elastic energy back into mechanical energy. The graph shows what happens when a material is subjected to stress and then the stress is removed. We assume that the material is not pushed into the plastic region and so it will return to its original shape.

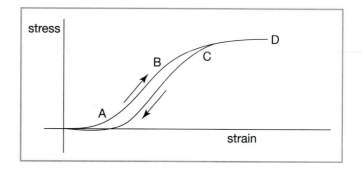

Along the upper curve, work is done on the material, and its internal energy ($\Delta E_{elastic}$) increases. Along the lower curve, the material's internal energy is decreasing. This loss of internal energy shows up in general as mechanical energy and thermal energy. The area between the two curves represents the amount of thermal energy that is generated. This graph is called a hysteresis loop (Enoka, 1994, p.105). In the case of an ideal spring, the two curves would exactly overlap; no thermal energy would be generated. For most biological materials, almost all of the internal energy shows up as thermal energy when the stress is removed. There are exceptions: elastin and resilin, as discussed earlier. For these materials, the area between the curves is very small, and so almost all of the internal energy shows up as mechanical energy when the applied stress is removed.

Since, as we pointed out earlier, there is very little elastin and no resilin in the human body, it seems unlikely that the storage of elastic energy and its subsequent conversion into mechanical energy play any significant role in human body dynamics. This is an area of active research, however, and new data are constantly being made available. There is some evidence that a malfunction of the elastin gene is associated with pediatric cardiac problems.

We have already introduced the idea of collisions (see the section on momentum). We showed that if a system (a collection of objects) is involved in a collision, the total momentum evaluated before the collision is equal to the total momentum evaluated after the collision. This idea is called conservation of momentum. We can now continue that discussion.

On the basis of the material discussed above, we would expect that the mechanical energy of a system would be less after a collision than before the collision. Thus, when a ball is dropped, it rebounds from the surface (ground, sidewalk, etc.) to a height that is less than its original height. The analysis of an actual collision, even a simple one that involves only two objects, is quite involved. As the objects collide, they deform, and a large part of the original mechanical energy is used in the deformations. Many objects, after undergoing a deformation, will regain their original shape (an elastic deformation). Upon regaining of the original shape, some (but not all) of the energy that was used in the deformation shows up as mechanical energy. This mechanical energy is represented by the kinetic energy of the objects after the collision.

EXAMPLE 3.13

The figure shows a collision between a foot and a football. The deformation of the ball is obvious. A significant amount of energy (transferred from the foot to the football) is required to produce such a large deformation. The foot loses kinetic energy (i.e., slows down) as it hits the ball, and the ball gains elastic energy. Some of this elastic energy will be dissipated as heat and noise, but the remainder shows up as kinetic energy of the football.

The indented part of the football cannot expand to the left because the kicker's foot is in the way. The deformation is removed by displacement of the football to the right. Thus the ball moves to the right; the elastic energy has been converted to kinetic energy.

EXAMPLE 3.14

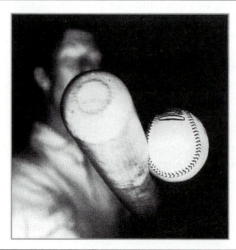

The figure shows a baseball as it collides with a bat. The partial flattening of the baseball is clear. As with the football, some of the elastic energy that is stored in the flattened baseball shows up as kinetic energy as the ball takes flight.

NUCLEAR ENERGY

By the end of the nineteenth century, kinetic energy, potential energy, chemical energy, thermal energy, and even electrical energy were accepted as physically real. In 1905, a new form of energy was predicted to exist, on the basis of theoretical rather than experimental evidence. Albert Einstein, then a 22-year-old patent reader in Switzerland, published a paper in which he made known his investigations into some unresolved questions about electricity and magnetism. He was able to present answers to these questions, but to arrive at his answers, he had made some radically new assumptions about the relationship between time and space. Before his work, it had generally been assumed that time and space were separate, each describing a totally different aspect of the world.

For Einstein to arrive at his results, he had assumed that time and space were not separate but were related to each other. In particular, he assumed that the speed of light would be the same for all experimenters no matter how fast they were moving, as long as they were not accelerating. This assumption led to several amazing results. One of these was that observers who were moving at different (constant) velocities would measure different time intervals between the same events. Nothing like this had ever been suggested before. A perhaps even more amazing result of his analysis was that energy was related to mass. The relationship was expressed by the equation

$$E = (c^2)(\Delta m)$$

It must be emphasized that this does not imply merely that the amount of energy (E) happens to be equal to an amount of mass (Δm) multiplied by the square of the speed of light (c^2). It states that energy is the same as mass. The quantity c^2 serves only to ensure that the units are the same on the two sides of the equation.

For example, if we write

$$\Delta KE_{joules} = \left(\frac{0.239 \text{ cal}}{1 \text{ J}}\right) TE_{calories}$$

we understand that the quantity in the parentheses serves only to adjust the units; kinetic energy and thermal energy are really the same. If we were to agree to measure them in the same units, we write

$$\Delta KE = TE$$

Thus Einstein was suggesting that mass is a form of energy just such as kinetic energy and thermal energy. This implied that, just as kinetic energy and thermal energy can be transformed into each other, so too can mass and kinetic energy.

Not only was the concept of mass transforming to energy difficult to accept, but also the magnitudes involved were almost beyond comprehension. Consider a pint of water, about 1 lb. How much kinetic energy would result from the transformation of this 1 lb of matter?

$$KE = (3 \times 10^8)^2 (1 \text{ lb}) \left(\frac{1 \text{ kg}}{2.21 \text{ lb}}\right)$$

$$= 4.1 \times 10^{16} \text{ J}$$

As difficult as it was for Einstein's ideas to be accepted in the early twentieth century, by the end of World War II, there was no question of their validity. The explosions of the atomic bombs proved unequivocally that mass could be transformed into other forms of energy, such as thermal energy. In the late 1940s and 1950s, many people expected that nuclear energy or atomic energy, as Einstein's predicted phenomenon had come to be known, would replace all other sources of electric, thermal, and mechanical energy. There would be no need for damming rivers, mining for coal, or the many polluting facilities wherein chemical energy was transformed to electrical energy. At that time, no one realized the long-term costs associated with nuclear waste disposal, environmental damage, and the dangers presented by possible malfunctions. We have found that these concerns are more important to our society than the societal gains listed above.

PROBLEM SET 8

8.1. A woman is eating a candy bar while backpacking up a mountain at constant speed. She ingests 750 kcal while her potential energy is increasing by 475 kcal. Her body produces 380 kcal of thermal energy due to the exertion. Calculate the change in her internal energy. (-105 kcal)

8.2. An 8000-kg San Francisco cable car is at rest on the top of a 30-m-high hill. Its brakes fail, and it "runs away" down the hill.
   a. Assume that there is no friction. How fast will it be moving as it gets to the bottom of the hill? (24.2 m/s)
   b. Suppose that the brakes fail only partially and that, as a result, the car is moving at 5 m/s at the bottom of the hill. How much thermal energy was produced in the partially working brakes? (2.25×10^6 J)

8.3. A pitcher throws a 115-g baseball at a speed of 100 mi/h ($= 44.7$ m/s). From the end of the windup to the instant when the ball is released, the ball moves 11.5 feet ($= 3.5$ m)
   a. Determine the change in kinetic energy of the ball. (115 J)
   b. How much work does the pitcher do on the ball? (115 J)
   c. Calculate the average force exerted by the pitcher on the ball. (32.8 N $= 7.4$ lb)

8.4. As a 250-g arrow is shot from a bow, the string pushes it with a force of 50 lb (= 222 N). The string is in contact with the arrow for the first 30 in. (0.76 m) of its flight.

a. How fast will the arrow be moving as it leaves the bow? (36.7 m/s)

The arrow eventually hits a target and penetrates it a distance of 4 in. It is known that the force of friction between the target and the arrow is 5 lb.

b. How much mechanical energy did the arrow "lose" during its flight through the air from the bow to the target? (167 J)

c. Where did this energy "go"?

8.5. The force of gravity exerted by the sun on the earth is very large. How much work does the sun do on the earth in 1 year? (Zero, EXPLAIN)

8.6. You lift 2 kg of potatoes from the floor to a counter that is 1.2 m high.
a. How much work do you do on the potatoes? (23.5 J)
b. How much work do the potatoes do on you? (−23.5 J)
c. Describe the transfer of energy. (You are transferring energy to the potatoes.)

8.7. Refer to Problem 8.3 and determine the rate at which the pitcher is doing work on the ball. (1.06 kW)

8.8. How much does it cost to operate a 100-W light bulb continuously for 1 week if the cost of electricity is $0.10 per kWh? ($1.68)

8.9. The BMR of a human being is about 100 W. Consider a person whose job it is to lift bags of potatoes (5 lb each) from the floor to a shelf that is 4 ft above the floor. How many bags have to be lifted each minute so that the person is working at the rate of 100 W? (222)

8.10. How many doughnuts are equivalent, in terms of chemical energy, to a gallon of gasoline? (212)

8.11. A typical residence uses 400 kWh of electricity per month. Use the following data and determine the cost of this amount of energy for each of the following sources of the energy:
a. Doughnuts at $0.35 each   ($891.59)
b. Gasoline at $1.45/gal   ($17.40)
c. Electricity at $0.11/kWh   ($44.00)
d. Coal at $166/ton   ($7.86)

8.12. Consider a woman who is resting. She is hooked up to an apparatus that measures the concentration of oxygen in the air that she inhales and in the air that she exhales. The inhaled air is 21% oxygen, and the exhaled air is 16% oxygen. The measurements show that she inhales (and exhales) 0.5 L of air with each breath. How many times per minute must she breathe (respiratory rate) for her body to satisfy her BMR requirements? (12)

8.13. We have discussed the BMR of a person in class and pointed out that it is about 100 W. This problem is intended to put this amount of power into perspective. Consider a 110-lb person who is doing push-ups. To do a push-up, she must raise her center of gravity by 8 inches. How many push-ups would she have to do per minute so that she would be working at the rate of 100 W? (60)

8.14. A 500-house development uses electricity at an average rate of 5 MW.
   a. How much energy does the development use each day? (4.3×10^{11} J)
   b. How much material would have to be converted in a nuclear reactor each day to supply this much energy? (4.8 mg)
   c. How many tons of coal would have to be burned to give the same amount of energy? (14 tons)

8.15. How many gallons of gasoline would have to be used to produce the same amount of energy as transforming the mass in 1 cup (0.5 lb) of water? (1.7×10^{8} gal)

8.16. Sketch the hysterisis curves for the following:
   a. A bow used in archery
   b. A ligament that has been strained into the plastic region

8.17. A reasonable estimate for the rate at which a person doing manual labor can work over an 8-hour day is 0.1 hp. A reasonable estimate for the amount of electrical energy used in a residence (none used for heat) over a month (thirty 24-hour days) is 400 kWh.
   a. How many people would have to be employed in the residence to supply an amount of work equal to the amount of electrical energy that was used? (an average of 8 people per 8-hour shift)
   b. Assume that the average wage earned by these people is $5 per hour and that electrical energy costs $0.11 per kWh, and determine the cost of the electricity and the wages earned by the people. (wages would be $320 per 8-hour shift compared to $0.50 for the electrical energy)

8.18. A 115-lb woman, while helping a friend move into a third-floor apartment, eats a doughnut. The apartment is 40 ft above the driveway. Her friend's belongings are packed into boxes that weigh 20 lb each. Assume that her efficiency when walking upstairs is 10%.
   a. How many boxes would she have to carry up to work off the energy gained from the doughnut? (If she carries 1 box per trip, she must move 8 boxes. If she carries 2 boxes per trip, she must move 14 boxes.)
   b. Assume that it takes her 3 hours to move these boxes, and calculate the rate (in hp) at which she is working. (7×10^{-4} hp)
   c. If all of the thermal energy that she developed remained within her body, by how much would her temperature rise? (2.8°C)

8.19. Why would one expect that the mechanical energy of a system would be less after a collision than before the collision?

THERMAL ENERGY AND EFFICIENCY

Heat is a form of energy that flows from one place to another because of a difference in the temperatures of the two locations. Referring to it as *thermal energy* will emphasize that it is a form of energy and not some other kind of stuff. However, since the use of the word "heat" is so ingrained, we shall use it and "thermal energy" interchangeably.

Microscopically, at the level of atoms and molecules, heat is the kinetic energy of atoms and molecules in random motion. We cannot observe these individual atoms and molecules moving randomly, but we can observe the effects of this motion, such as changes in size, shape, and physical state. We commonly use the term "heat" to explain these observations rather than "random molecular kinetic energy" because it is more familiar and hence, seems simpler.

We can now understand the energy balance in the body and the roles played by diet, metabolism, and exercise. When we eat food and digest it, some of the available energy is stored, and some shows up as heat. When we exert ourselves, we use some of this stored energy. If we take in energy faster than we use it, we gain weight. To lose weight, we must use energy faster than we are taking it in. This can be accomplished by taking in less energy (dieting) or by using more of it (exercising).

If I am involved in a physical activity, I will be using up stored chemical energy, and as a result of the activity, mechanical energy will become apparent. Later we will discuss the need of the body to maintain a relatively constant temperature, but for the present, we may ignore this. Assume, for example, that the activity is taking place in a very warm environment so that the body is kept warm externally. Ideally, all of the chemical energy represented by some fat in my body could be transformed into mechanical energy, for example, by my riding a bicycle, working, or running. Unfortunately, this is not the case. Whenever energy is changed from one form to another, such as chemical energy to kinetic energy, some of it always shows up as heat. Usually, this amount of energy that has been transformed into heat is wasted. This happens within the body, when chemical energy is transformed to mechanical energy in the muscles. It also happens outside the body. For example, if I am riding my bike on a level road ($\Delta PE = 0$) at constant speed ($\Delta KE = 0$), all of the chemical energy that my body is consuming shows up as heat produced within my body and by friction between the bicycle and the air (and the road).

Using this terminology, we may restate conservation of energy as

$$\text{what is consumed} = \text{what is accomplished} + \text{what is wasted}$$

So, in general, the input energy results in some useful output and some that is wasted. For example, the electrical energy that is consumed by a light bulb results in some useful output (light, or radiant energy) and some wasted energy (heat). The advantage of fluorescent lights is that much more of the electrical energy shows up as light, and correspondingly, less is wasted as heat. Thus although a 14-W fluorescent light consumes only about 1/7 as much electrical energy as a 100-W ordinary (incandescent) bulb, it produces about the same amount of light. Of course, it produces much less heat and is therefore cooler to the touch.

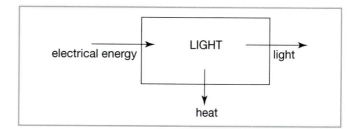

For an automobile, the useful output is the mechanical energy of the car, and once again, the wasted energy is the heat that is produced both within the motor and by friction with the road surface and the air.

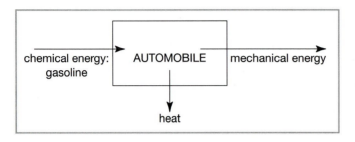

In the case of the human body, the input is metabolized fat, sugar, and so on, and the desired output might be the increase in potential energy associated with climbing a mountain.

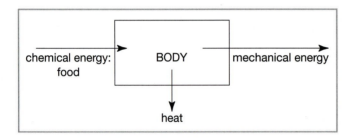

The ratio of useful energy output to energy input or alternatively the ratio of useful power output to power input is called the **efficiency**:

$$e = \frac{\text{what is accomplished during the activity}}{\text{what is consumed during the activity}}$$

This basic definition can be rewritten as:

$$e = \frac{\text{what is consumed during the activity} - \text{what is wasted during the activity}}{\text{what is consumed during the activity}}$$

$$= 1 - \frac{\text{what is wasted during the activity}}{\text{what is consumed during the activity}}$$

From this expression, we see that the maximum value of efficiency is 1 (if nothing is wasted) but that it is, more realistically, less than 1.

It is important to note that both the numerator and the denominator of the defining equation must have the same units, such as joules/joules or watts/watts or lb ft/lb ft.

The chart lists the efficiencies of various types of machines and the human body under various activities.

ACTIVITIES	%
Swimming	2
Steam engine	10–15
Gasoline automobile	14
Bicycling	20
Photosynthesis	23–43
Skeletal muscle	25
Gasoline engine	38
Heart	40
Fuel cell (Pt in NaOH)	50
Diesel engine	56
Propulsion of a water beetle	84

The part of theoretical physics that deals with heat and energy is **thermodynamics.** Just as Newton's three laws form the basis of mechanics, there are three concepts (laws) that are basic to thermodynamics. The first law is conservation of energy, which we have already discussed. The second law has several forms, but one of them deals with the efficiency whereby heat may be transformed to mechanical energy. In the early nineteenth century, a French mathematician/engineer, Sadi Carnot, derived the following formula. It expresses the maximum efficiency for any heat engine that takes in heat at a high temperature and gives off heat at a lower temperature; during this process work is done. The formula for this efficiency is

$$e_{\text{Carnot}} = \frac{T_{\text{hot}} - T_{\text{cold}}}{T_{\text{hot}}}$$

where the temperatures must be in kelvin.

EXAMPLE 3.15

Although a biological entity is not a heat engine, it is interesting to compare our muscles to an ideal heat engine. If we assume that T_{cold} is represented by the environment (68°F = 20°C = 293K), our muscles would have to work at a temperature of about 393K (120°C = 248°F) to achieve an efficiency of 25%. However, our muscles actually work at a temperature of 37°C. Our ability to achieve a high level of efficiency at such a low operating temperature (compare 37°C to 120°C) is due to the wondrous contribution made by mitochondria to our metabolic processes. If the mitochondria are not functioning properly, our muscles will not be capable of efficiently converting stored chemical energy into mechanical energy. An example of this is the unfortunate situation faced by Greg LeMond (*New York Times*, Dec. 3, 1994, p. 31), who for many years was the best U.S. bicycle road racer and three-time winner of the Tour de France. He was found to be suffering from mitochondrial myopathy—basically, dysfunctional mitochondria. His body is no longer capable of converting stored chemical energy into mechanical energy fast enough for him to race.

EXAMPLE 3.16

In an earlier section, we determined that a woman would have to do almost 7000 sit-ups to burn off the energy that she gained by eating 1/4 pint of ice cream. We can now carry out a more realistic analysis.

Solution

A reasonable value for her efficiency when exercising is 10%. Thus the gain in potential energy represents only 10% of her consumed energy. We already know that the potential energy increases by 121 J per sit-up. So if she does n sit-ups, the potential energy increases by $121n$, and we have:

$$e = \frac{\text{what is accomplished during the activity}}{\text{what is consumed during the activity}}$$

$$0.1 = \frac{121n \text{ J}}{841 \text{ kJ}}$$

$$n \sim 700 \text{ sit-ups}$$

This is a more realistic result but is still not a reasonable way to deal with eating the ice cream.

EXAMPLE 3.17

While using a biceps curl exercise machine, a person raises a 5-kg load though a distance of 35 cm in 0.25 s. While doing so, her biceps muscle contracts (shortens) by 4 cm. Her efficiency during the exercise is 10%.

1. Calculate her useful power output during this exercise.
2. How many repetitions must she do to burn off the energy that she took in while eating six chocolate chip cookies?
3. How much thermal energy does her body produce during the workout?
4. How much force is exerted by her biceps?

Solution

We must recognize that the object of interest (her body) is doing work on the load because the height of the load is increasing as a result of a force that her body applies. Her body is converting stored chemical energy into work and thermal energy. Since we are considering more than one form of energy, we will begin with the conservation of energy equation. Only three terms will play a role:

$$0 = W_{\text{by her body}} + TE_{\text{leaving her body}} + \Delta U_{\text{of her body}}$$

She does work each time she lifts the load. She applies a force (equal to the weight of the load) upward, and the load is displaced upward; thus θ is 0°. The amount of work she does during each repetition is

$$W_{\text{per repetition}} = m_{\text{load}} g \Delta h_{\text{load}}$$
$$= (5)(9.8)(0.35)(\cos 0°)$$
$$= 17.15 \text{ J}$$

$$P_{\text{output}} = \frac{W_{\text{output}}}{\Delta t}$$
$$= \frac{17.15 \text{ J}}{0.25 \text{ s}} = 68.6 \text{ W} = 0.09 \text{ hp}$$

If she performs n repetitions, the amount of work she does is

$$W = nm_{load}g\Delta h_{load}$$
$$= n(17.15 \text{ J})$$

She uses up energy that is equivalent to six chocolate chip cookies:

$$\Delta U = -(6)(35 \text{ kcal})$$

Here we notice a problem with units (we have kcal mixed with m and kg). Let us agree to stay with the SI system, so we have to change kcal to J:

$$\Delta U = -879 \text{ kJ}$$

Substituting into the conservation of energy equation, we have:

$$0 = n(17.15 \text{ J}) + TE_{\text{leaving her body}} - 879 \text{ kJ}$$

Notice that we have two unknowns: n and TE; therefore we need another equation. The efficiency provides the needed equation:

$$e = \frac{\text{what is accomplished during the activity}}{\text{what is consumed during the activity}}$$

$$e = \frac{\text{work done on the load}}{\text{the amount of energy used by her body}}$$

$$0.1 = \frac{n(17.15 \text{ J})}{879 \text{ KJ}}$$

We can now evaluate n. **She has to carry out 5125 repetitions. Obviously, this is not a realistic way to compensate for eating some cookies.** Now that we have evaluated n, we can determine how much thermal energy her body produced. Substituting into the conservation of energy equation, we have:

$$0 = (5125)(17.15 \text{ J}) + TE_{\text{leaving her body}} - 879 \text{ kJ}$$

We see that her body produced 791 kJ of thermal energy. We shall see later that it is very important that her body can get rid of this thermal energy.

To determine the amount of force exerted by her biceps, we may assume that the amount of work done by her biceps is equal to the amount of work done on the load:

$$|W_{\text{by biceps}}| = |F_{\text{biceps}}||\Delta L_{\text{biceps}}|$$

$$\left|\frac{|17.15|}{2}\right| = |F_{\text{biceps}}||0.04|$$

The left-hand side was divided by 2 because while both arms are used to lift the load, the calculation of the force exerted by the biceps refers to a single arm.

THERMAL ENERGY RELATED TO CHANGE IN TEMPERATURE

In the section on internal energy, we saw that one form of internal energy is thermal energy. Although the amount of thermal energy within an object is not directly observable or measurable, we can detect and determine the change in this quantity. When the amount of thermal energy within an object changes, there are two observable effects:

1. a change in temperature
2. a change in physical state

We will first discuss temperature. Although the idea of temperature seems very familiar, it is not straightforward at all. Our concept of temperature is based on our physiological perception of "hot" and "cold." These perceptions are not quantitative and are not even unequivocal. When first stepping into a bath, I may perceive that the water is hot. After a few minutes, I may perceive the water to be warm or even cool. A thermometer in the water would show that although my perception had changed rapidly, the temperature had changed very slowly if at all. If I go to the beach in August and the water temperature is 75°F, I might feel that it is too cold for swimming. If I were to go to the beach in January and the water temperature was 75°F, I would feel that it was very warm—**the same temperatures but very different perceptions.**

We cannot depend on our perceptions but must depend on an instrument, a **thermometer.** A thermometer is a device that has a measurable change that depends in a reliable way on temperature. Thermometers may be based on a change in volume of a fluid (common fever thermometer), change of electrical resistance (thermistor), or even a change in color (liquid crystal thermometer). No matter what the physical basis for the operation of the thermometer, we can get quantitative information about the temperature.

There are three temperature scales in common use. These are **Fahrenheit, Celsius** (called "centigrade" before 1948), and **Kelvin** (also called "absolute"). The Fahrenheit and Celsius scales are based on setting two fixed temperatures and then dividing the interval between them into 100 divisions. In the case of the Fahrenheit scale, the two fixed points are represented by a mixture of ice/water/salt (arbitrarily chosen to be at 0°F) and human body temperature (originally chosen to be at 100°F.) The Celsius scale uses ice/distilled water for 0°C and water/steam for 100°C.

During the eighteenth century, it was experimentally determined that as the temperature of a fixed amount of gas was lowered (at constant pressure), the volume decreased. The decrease in volume was determined to be 1/273 of the original volume for each °C that the temperature was lowered below 0°C. On the basis of these data, it was thought that if the temperature were lowered to 273°C below 0°C, the volume might go to zero. This temperature was therefore selected to be absolute zero. A temperature scale based on absolute zero rather than on two arbitrary fixed points is called an absolute scale, and the Kelvin scale is so based. In general, one may use any temperature scale, but there are certain conventions. In the United States, the commonly used scale is Fahrenheit. In the rest of the world and in the scientific and medical communities, the commonly used scale is Celsius.

In dealing with problems involving gases and/or thermal (blackbody) radiation, the Kelvin scale *must* be used.

Temperatures may be converted from one to another of these scales by the use of the following formulas:

$$(°F) = \frac{9}{5}(°C) + 32$$

$$(°C) = K - 273$$

SAMPLE TEMPERATURES	K
Center of hydrogen bomb explosion	10^8
Highest temperature attained in a lab	6×10^7
Center of the sun	1.5×10^7
Surface of the sun	4.5×10^3
Acetylene flame	2900
Combustion chamber of a gasoline engine	2770
Melting point of iron	1800
Melting point of lead	600
Human internal	310
Liquid oxygen	87
Liquid nitrogen	77
Liquid hydrogen	20
Liquid helium	4.2
Interstellar space	3
He^4 (under pumping)	1
Mixture He^4 and He^3	0.002
Lowest temperature attained in a lab	3×10^{-8}

The original (early nineteenth century) work on the concept of temperature was based on the fluid (caloric) model of heat. When it was later established that heat is a form of energy, the modern concept of temperature started to develop. In the late nineteenth and early twentieth centuries, the atomic theory became widely accepted, and this was applied to the question of heat and temperature. When the atomic theory was applied to ideal gases, the following formula was derived:

$$T = \frac{m(v^2)_{average}}{3k}$$

where T is the temperature in Kelvin, m is the mass of each of the particles (atoms or molecules) that make up the gas, $(v^2)_{average}$ is the average of the squares of the random velocities of all of these particles, and k is the Boltzmann constant $(= 1.38 \times 10^{-23}$ SI). It is interesting to note that if the gas were to be at a temperature of 0 K, the average random velocity of its atoms or molecules should be zero. This is what led to the idea that all motion ceases at absolute zero.

It is important to note that it is the random velocities of the particles that are important, not the organized or coordinated velocities. Thus when a wind is blowing, the temperature of the air is not thereby higher. The wind is the organized or coordinated motion of the molecules that make up the air. The temperature of the air does not depend on this type of motion but rather on the random motion of the molecules.

Although the relationship between temperature and average random velocity of constituents was derived for ideal gases, the concept was carried over to other systems, that is, liquids and solids. The formula does not work very well for materials other than ideal gases and must be modified. This modification was provided by the application of quantum mechanics to the understanding of heat and temperature in a solid material.

The relationship above shows that as the temperature of an object increases, the average random velocity of its constituents increases. This idea provides a starting point in the effort to understand some phenomena. For example, most objects get larger as their temperatures increase. We can imagine that as its temperature increases, its constituents are moving faster, randomly. Thus they take up more room, and the object itself takes up more room. A similar line of reasoning can be used to explain changes in state (to be discussed later).

As we pointed out earlier, the caloric theory was developed by the Scottish chemist Joseph Black. Although this model was later shown to be incorrect, it was based on what were, for the time, very precise measurements. Black tried to experimentally compare the amounts of heat that were involved in melting ice, heating water, and boiling water. He then tried to imagine a model for heat that would be consistent with his measurements. Although the following material does not describe exactly what Black did, it is close and easier to understand.

Imagine 454-g of water (about 2 cups) that has been frozen into a block of ice. The ice is then cooled to −15°C. This piece of ice is placed in an insulated container with a thermometer. The container is heated at the rate of 100 W, and data are recorded regarding the temperature of the contents and any noticeable activity. The graph represents such data.

The data show that for the first 2.4 minutes, the temperature of the ice increases steadily until it reaches 0°C (b). Then, during the next 25 minutes, the temperature does not change, but it is observed that the ice melts. At the end of the 25 minutes (c), all of the ice has melted, and then the temperature of the resulting water is observed to steadily increase for the next 32 minutes, until it reaches 100°C; at that point (d), the water starts to boil. During the next 171 minutes (almost 3 hours), the water continues to boil (i.e., changing to steam). The temperature does not change during this time. After all of the water has turned into steam (e), the temperature increases steadily again.

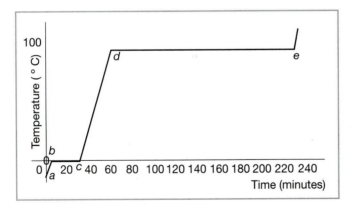

It is important to note the following:
- The temperature increases linearly between a and b, between c and d, and beyond e.
- The slopes between a and b and beyond e are approximately the same and are about twice as large as the slope between c and d.
- The temperature does not change between b and c or between d and e even though the container is being steadily heated.

The first two of these observations lead to the concept of specific heat (described below), and the third observation leads to the concept of latent heat (described later).

If an object gains or loses a quantity of thermal energy and there is an accompanying change in temperature, the following relation holds:

$$TE_T = mc\Delta T$$

where m is the mass of the object, ΔT is the change in temperature, c is a quantity called the **specific heat**, and TE_T is the amount of thermal energy transferred. If the body gains thermal energy, ΔT will be positive. If the body loses thermal energy, ΔT will be negative. The following table lists the specific heats of several common materials:

MATERIAL	SPECIFIC HEAT $\left(\dfrac{\text{calories}}{\text{gram °C}}\right)$
Aluminum	0.214
Copper	0.092
Silver	0.056
Steel	0.11
Ice $(-10°C)$	0.53
Ethanol	0.581
Mineral oil	0.5
Glass	0.2
Granite	0.19
Pure water	1.0
Seawater	0.93
Human body (average)	0.83

Notice that water has the highest specific heat of the common substances. This will turn out to be very important in understanding heat and temperature in our world.

EXAMPLE 3.18

In the section on efficiency, we showed that a 115-lb woman using a biceps machine produced 791 kJ of thermal energy while exercising. Assume that all of this thermal energy remained within her, and determine her increase in body temperature.

Solution

The relation between thermal energy and temperature is the specific heat formula:

$$(791 \text{ kJ})\left(\frac{0.239 \text{ cal}}{1 \text{ J}}\right)\left(\frac{1000 \text{ J}}{1 \text{ kJ}}\right) = (115 \text{ lb})\left(\frac{1 \text{ kg}}{2.21 \text{ lb}}\right)\left(0.83 \frac{\text{cal}}{\text{gr °C}}\right)(\Delta T)\left(\frac{1000 \text{ gr}}{1 \text{ kg}}\right)$$

$$TE_T = mc\Delta T$$

$$\Delta T = 4.4°C$$

This rise in temperature would probably be fatal, so the thermal energy that her body produces must be removed from it.

LATENT HEAT

As we saw in the previous section, it is quite possible for something to be heated without an accompanying change in temperature. We also saw that when this happens, there is an accompanying change in physical state (solid, liquid, gas) of whatever is being heated. These changes in state are called fusion (involving solid and liquid), vaporization (involving liquid and gas), and sublimation (involving solid and gas).

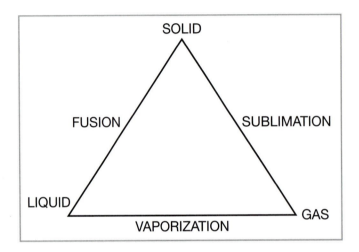

The amount of thermal energy involved in the change of state, TE_L, when a mass (m) of material changes state may be determined from the following definition:

$$TE_L = (H_L)(m)$$

where H_L (called the latent heat) depends on the specific material and process and is determined from the following table. The temperatures at which specific transformations occur are also shown in the table. For our purposes, the latent heats of water will play a primary role.

The table shows the melting point (the temperature at which fusion occurs at a pressure of 1 atmosphere, the boiling point (the temperature at which vaporization occurs at a pressure of 1 atmosphere) and the respective latent heats of various substances.

SUBSTANCE	MP(K)	H_F (J/g)	BP(K)	H_V (J/g)
Helium	3.5	5.23	4.126	20.9
Nitrogen	63.18	25.5	77.34	201
Ethyl alcohol	159	104.2	351	854
Mercury	234	11.8	630	272
Water	273.15	335	373.15	2256 (2428 @ 37°C)
Lead	600.5	24.5	2023	871

Notice that there are two values for the heat of vaporization (H_v) of water. The value 2256 J/g is to be used when water is changing between the fluid and vapor states at its normal boiling point (100°C). However, when water evaporates from the skin, its temperature is not even close to 100°C; its temperature is closer to the normal body temperature, 37°C. **Therefore, any analysis that involves the evaporation of sweat from the body should use 2428 J/g as H_V for water.**

The process of evaporation of sweat is a major factor in the ability of the body to regulate its temperature. For brief periods of time, the body may produce sweat at the rate of up to 4 L/h. A more typical value for prolonged sweating is 1 L/h. (Davidovits, p. 142. See Bibliography). **Only sweat that actually evaporates is useful in cooling the skin; sweat that drips off the body plays a very small role in temperature regulation.**

This is very important, and if the process of sweating is somehow blocked, a dangerous situation may result. For example, a common practice of high school wrestlers is to try to lose weight, just before the weigh-in for a match. Wrestlers are paired off on the basis of weight not age or strength. So an older or stronger wrestler who can lose weight may acquire an advantage by competing against a younger or weaker opponent. The fastest way to lose weight is to sweat it off. This is accomplished by putting on a rubber or nylon suit and then exercising. These materials are effectively waterproof, and so the water vapor produced by evaporating sweat stays close to the body. This raises the humidity in the immediate environment of the body. In a very humid environment, the rate of evaporation decreases dramatically. As the body's internal temperature increases (because of the exertion), sweat is produced. However, since it cannot evaporate, the body's internal temperature does not decrease, and so more sweat is produced. The goal of losing weight is achieved, but the loss of a great deal of fluid may lead to dehydration, a potentially fatal condition.

EXAMPLE 3.19

A 170-lb man may lose 6 lb of water while riding a bicycle on a very hot day.

1. How much thermal energy did his body lose while evaporating this much sweat?
2. If all of this thermal energy had remained within his body, what would have been the result?

Solution

1. The amount of thermal energy involved in a change of state is given by

$$TE_L = (H_L)(m)$$

$$TE_{vaporization} = \left(2428 \frac{J}{g}\right)\left(6 \text{ lb} \frac{1000 \text{ g}}{2.21 \text{ lb}}\right)$$

Thus he would have lost 6592 kJ = 6.59 MJ.

2. The relation between thermal energy and change in temperature is given by

$$TE_T = mc \, \Delta T$$

$$6592 \text{ kJ} = (170 \text{ lb})\left(0.83 \frac{cal}{g°C}\right)(\Delta T)\left(\frac{1 \text{ J}}{0.239 \text{ cal}}\right)\left(\frac{1 \text{ kJ}}{1000 \text{ J}}\right)\left(\frac{1000 \text{ g}}{2.21 \text{ lb}}\right)$$

Thus his temperature would increase by 24°C, and he would have been long dead. This shows how important sweating is during exercise.

EXAMPLE 3.20

Two ice cubes (3 g each) at 20°F ($-6.67°C$) are placed into an insulated copper cup (40 gram) that contains 113 grams of water at 54°F (12.2°C). Determine the final temperature of the cup of water.

Solution

We will assume that since the cup is insulated, there is no exchange of thermal energy with the environment. Therefore the total amount of thermal energy must not change. Any increase of thermal energy by the ice must be compensated for by a loss of thermal energy by the warm water and cup.

$$0 = TE_{T,ice} + TE_{T,water} + TE_{T,cup}$$

$$0 = (6 \text{ g}) \left(0.53 \, \frac{cal}{g°C} \right) (T_{final} - (-6.67°C)) + (113 \text{ g}) \left(1 \, \frac{cal}{g°C} \right) (T_{final} - 12.2°C)$$

$$+ (40 \text{ g}) \left(0.092 \, \frac{cal}{g°C} \right) (T_{final} - 12.2°C)$$

$$T_{final} = 11.7°C$$

We have made the usual assumptions, that is, that the cup and the water had the same initial temperature and that the entire system (cup, water, and ice) has the same final temperature. We have also used the general sign convention, whereby:

$$\Delta T = T_{final} - T_{initial}$$

Although the result ($T_{final} = 11.7°C$) is mathematically correct, it is physically impossible. It suggests that the final temperature of the ice is 11.7°C. However, under normal conditions, **the temperature of ice cannot exceed 0°C.**

We have not paid attention to the sequence of events as time passes. As the temperature of the warm water and cup decreases, thermal energy is produced. This thermal energy cannot escape to the environment (because the cup is insulated), and so it must be absorbed by the ice. As the ice absorbs thermal energy, several changes take place *in sequence.* First the temperature of the ice increases from its initial value. This goes on until the temperature of all of the ice reaches 0°C. At this point, ice starts to melt, the temperature remaining at 0°C. After all of the ice has melted, the temperature of the resulting water begins to increase from 0°C. Of course, while all of this has been happening, the temperature of the warm water and cup has been decreasing. Eventually, the temperature of the warming ice water will be the same as the temperature of the cooling warm water and cup. This will be the final temperature: thermal equilibrium.

The amount of thermal energy that the ice must absorb in order to melt is given by

$$TE_F = (H_F)(m)$$

$$= \left(335 \, \frac{J}{g} \right) \left(\frac{0.239 \text{ cal}}{1 \text{ J}} \right) (6 \text{ g})$$

$$= 480.4 \text{ cal}$$

If this should occur, the cup would contain the original 113 g of water and 6 g of melted ice (let's call it ice water to distinguish it from the original water). After all of the ice has melted (because it absorbed thermal energy from the warm water and cup), the ice water would warm up (as it absorbed thermal energy from the warm water and cup):

$$0 = TE_{T, ice} + TE_{T, water} + TE_{T,cup} + TE_{fusion, water} + TE_{T, warming\ ice\ water}$$

$$0 = m_{ice}c_{ice}\Delta T_{ice} + m_{water}c_{water}\Delta T_{water} + m_{cup}c_{cup}\Delta T_{cup}$$
$$+ H_{fusion, water}m_{melted, ice} + m_{icewater}c_{water}\Delta T_{icewater}$$

$$0 = (6\ g)\left(0.53\ \frac{cal}{g°C}\right)(T_{final} - (-6.67°C)) + (113\ g)\left(1\ \frac{cal}{g°C}\right)(T_{final} - 12.2°C)$$

$$+ (40\ g)\left(0.092\ \frac{cal}{g°C}\right)(T_{final} - 12.2°C) + 480.4\ cal + (6\ g)\left(1\ \frac{cal}{g°C}\right)(T_{final} - 0°C)$$

$$T_{final} = 7.3°C = 45°F$$

PROBLEM SET 9

9.1. Normal human body temperature is taken to be 98.6°F. Express this temperature in kelvin. (310 K)

9.2. What temperature on the Fahrenheit scale has the same numerical value on the Celsius scale? ($-40°F$)

9.3. The boiling point of liquid helium is taken to be 4.2 K. Express this temperature on the Fahrenheit scale. ($-451.84°F$)

9.4. The title of the book and film, *Fahrenheit 451* by Ray Bradbury refers to the combustion temperature of paper. What might a European author have entitled the book? (*Celsius 233*)

9.5. How much thermal energy must be added to a 150-lb person for the person's internal temperature to rise from 34°C to 37°C? (169.5 kcal)

9.6. Consider a 120-lb person who is rubbing her hands together because they are cold. It is known that the average force of friction exerted by one hand on the other is 5 lb and the average distance the hands slide while being rubbed together is 6 in.
   a. How much thermal energy is required to warm her hands up by 5°C ? (2.71 kcal)
   b. Where does the thermal energy that is warming up her hands come from?
   c. How many times does she have to rub her hands together to raise their temperature? (3347)
   d. Does this seem a reasonable way of warming up her hands? Why or why not?
   e. If this process (part c) is 12% efficient, how much energy did the body use while warming up her hands? (22.6 kcal)
   f. If none of this energy is lost to the environment, what will be the effect on her body? (average body temperature will increase by 0.5°C)
   g. How many grams of sugar did she metabolize while rubbing her hands together? (6 g)

9.7.   When a scalpel is sterilized, its temperature may rise to 150°C. How much thermal energy must be extracted from a 30-g steel scalpel if its temperature is to be lowered from 150°C to 37°C? (373 cal)

9.8.   The useful power output of Bryan Allen (150 lb), who flew the first human-powered airplane across the English Channel in 2.5 hours on June 12, 1979, was 350 W. He powered the plane by pedaling a bicycle-type mechanism that was connected to the propeller.
   a. How much thermal energy did his body produce during the flight? (12.6 Mcal)
   b. If all of this thermal energy stayed in his body, by how much would his temperature have changed? (53.4°C)
   c. How much energy did his body use up during this time? (1.58 Mcal)
   d. Assume that all of this energy came from the oxidation of glucose, and determine how much glucose was used. (1.1 kg)

9.9.   After strenuous exercise, a 90-kg person has a body temperature of 40°C and is giving off thermal energy at the rate of 50 cal/s.
   a. How much thermal energy must this person lose for the body temperature to reach 37°C? (224 kcal)
   b. How long will it take for the body to reach this temperature? (75 min)
   c. If it is assumed that all of this thermal energy loss is due to evaporation of perspiration, how much water will be evaporated? (386 g)

9.10.   Ten grams of steam at 100°C is added to 100 g of water that is at an initial temperature of 25°C. The water is in an insulated 50-g copper cup. What is the final temperature of the system? (80.4°C)

9.11.   Consider a person who has a caloric intake of 2000 kcal. Assume that all of this energy either goes into work done by the person or is lost to the environment in the form of thermal energy. The person is working at a task for which the efficiency is 15%. Assume that all of the thermal energy lost is due to evaporating perspiration, and determine how much water is evaporated. (2.9 kg)

9.12.   A large power plant converts 1000 kg of water to steam every second to run its generators.
   a. How much thermal energy is needed each second to raise the temperature of the incoming water (at 15°C) to the boiling point? (85 Mcal)
   b. How much thermal energy is needed each second to turn this water into steam? (540 Mcal)
   c. Assuming that the efficiency of the plant is 15%, calculate the number of tons of coal used by the plant each day (24 hours). (49 kilotons)

9.13.   At what rate (in kcal/h) is a person losing thermal energy if he is evaporating 1 L of sweat per hour? (580 kcal/h)

9.14.   According to the American Heart Association, a 150-lb person consumes energy at the rate of 12.2 kcal/min when jogging on a level surface at a constant speed of 6 mi/h.
   a. How much thermal energy does this person produce during a 30-min run? (366 kcal)
   b. If all of this thermal energy stayed in the person's body, by how much would his temperature rise? (6.5°C)

     c. If this energy were to be produced by the oxidation of glucose, how much glucose would be consumed? (96 g)

     d. If all of the thermal energy were to leave his body by the evaporation of sweat, how much water weight would he lose? (631 g)

9.15. Refer to the discussion of Bryan Allen's flying a human-powered plane across the English Channel. How much water would he have to have lost through sweating in order that his body temperature did not increase? (11.5 lb)

9.16. In the section on specific heat, we showed that a 115-lb woman who was exercising with a biceps curl machine would produce 791 kJ of thermal energy and that as a result, her internal temperature would increase by 4.4°C. How much sweat would she have to evaporate to remove this thermal energy from her body? (326 g = 0.72 lb)

9.17. Heat stroke is evidenced by the absence of sweating by a clearly overheated person. Why is this considered to be dangerous?

9.18. The BMR (the rate at which the body uses energy when it is completely at rest) of a human being is 100 W. How many pounds of water would a person have to evaporate during an 8-hour sleep period to remove the resulting thermal energy? (2.6 lb)

9.19. When a person requires a simple surgical procedure on the skin, a spray is often used as a local anesthetic. Explain how such a spray could lessen the pain.

9.20. Explain why there is less variation of temperature in a region that is near a large body of water (e.g., the ocean) than in a region that is far inland.

THERMAL ENERGY TRANSFER

Humans are better suited to dealing with high-temperature environments than with low-temperature ones. There are nerves in the skin that send messages of pain when the temperature gets too high. There are no corresponding nerves identifying low temperature. In fact, if the temperature becomes too low, the pain nerves stop functioning, and the person may be unaware that a portion of the body is suffering from severe cold, as with frostbite. If we get too hot, we automatically try to get away from the heat to avoid the pain. If we get too cold, we automatically go to sleep and perhaps perish from **hypothermia.** Our automatic body defenses against low temperature are either potentially damaging, such as going to sleep or decreased blood circulation to the extremities, or relatively ineffective, such as shivering or getting gooseflesh. The corresponding reflex actions dealing with high temperature, such as sweating, increased blood circulation to the skin, and avoiding the sun to lessen the pain of sunburn, are effective and are not damaging.

    We must be very concerned with the flow of thermal energy from one place to another. It is sometimes necessary for thermal energy that is generated within our bodies to be transferred to the skin and then away. On the other hand, it is sometimes necessary that this transfer away be severely restricted. Therefore we must deal not only with the generation of thermal energy within the body but also with its transfer, both within the body and in its exchange with the environment.

EXAMPLE 3.21

Consider a 115-lb woman who is riding a bicycle at a moderate speed.
1. At what rate is she consuming energy?
2. At what rate is she producing thermal energy?
3. If all of this thermal energy were to remain inside her body, at what rate would her temperature increase?
4. What is the implication of this result?

Solution

1. Her body weight corresponds to 52.2 kg, so, by using the material discussed on page 152, she is consuming energy at the rate of 191 kcal/h.
2. A reasonable assumption for the efficiency of the body when cycling is 20%; therefore 80% of her energy consumption shows up as thermal energy. Using the material discussed on page 168, we see that she is thus producing thermal energy at the rate of 153 kcal/h.
3. The relation between thermal energy and temperature is given by the specific heat equation on page 175:

$$TE = mc\Delta T$$

$$\frac{TE}{\Delta t} = mc\frac{\Delta T}{\Delta t}$$

$$153\,\frac{\text{kcal}}{\text{h}} = (52.2\text{ kg})\left(0.83\,\frac{\text{kcal}}{\text{kg °C}}\right)\left(\frac{\Delta T}{\Delta t}\right)$$

From this equation, the rate of the change in temperature may be calculated:

$$\frac{\Delta T}{\Delta t} = 3.53\,\frac{\text{°C}}{\text{h}}$$

4. **As we discussed earlier, a rise of internal human body temperature can be very dangerous, and this woman would be seriously at risk in much less than 1 hour. This means that there must be a very efficient way for her body to get rid of this thermal energy.**

Blood, which is 80% water, has a high specific heat and a high latent heat of vaporization. Thus blood can absorb a great deal of thermal energy without vaporizing or even its temperature increasing excessively. As cool blood passes through high-temperature parts of the body (e.g., the liver and heart), it absorbs a lot of thermal energy, and its temperature rises. When it passes through a low-temperature region of the body (e.g., the skin), it loses thermal energy, and its temperature drops. The cool blood may now circulate to the higher-temperature interior, and the process begins anew. Thus thermal energy is continuously transferred from hotter to cooler parts of your body. Of course, when the thermal energy gets to your skin, it may easily leave the body.

We humans have a significant advantage over some other warm-blooded animals such as cats and dogs. We can sweat, and they can't. Your body can produce large amounts of water on your skin, and then this water can evaporate. For water (sweat) on the skin to evaporate, it must acquire a large amount of thermal energy. It absorbs this thermal energy from the skin. Thus the thermal energy that the blood transferred to the skin is now transferred from the body to the environment via the evaporation of sweat. The process of sweating plays a major role in the body's ability to regulate its internal temperature. A person may lose as much as 1 lb of water in nonobvious perspiration each day. The thermal energy loss represented by this evaporation represents on the order of 10% of a person's normal energy intake per day.

The normal flow of thermal energy is always from a higher temperature region to a lower-temperature region. The body's normal internal temperature is 37°C. In a cold environment, the skin temperature might be 27°C, and therefore thermal energy would flow from the interior of the body outward. The temperature of the interior would drop, and hypothermia might result. In a hot environment, the skin temperature might be 40°C. Thermal energy would then flow from the environment into the core of the body, causing excessive overheating. It is thus necessary for the body to be able to regulate its temperature by dealing with excessive flow of thermal energy both into the body and out from the body.

There are three ways in which thermal energy may be transferred: convection, conduction, and radiation.

Convection is thermal energy transfer that is accomplished by a medium moving from one place to another. Common examples of convection include the following:

Coolant circulating in an automobile

Hot air or hot water circulating through a home heating system

The Gulf Stream, a flow of warm water from the equator up along the coast of North America and then across the North Atlantic to Ireland

Blood circulating in our bodies

Air moving across exposed skin

In each of these examples, a physical fluid (water, antifreeze, air, blood) moves from a high-temperature location to a low-temperature location, carrying thermal energy with it. Since the moving material carries the thermal energy with it, the rate at which the thermal energy is transferred depends on the amount of moving material and its speed. For example, the amount of blood that travels through a vein and its speed depend on the diameter of the vein. A larger vein will carry more blood, faster. So when you are overheated, your skin may seem more red than otherwise. This is because the veins just under the skin have dilated (increased in diameter), the rate of blood flow through them increases, and hence the rate of convection of heat within your body increases.

Blood flow represents the major mechanism of thermal energy transfer within the body. The blood picks up thermal energy as it passes through or near high-temperature organs such as the heart and liver. It then loses thermal energy as it passes near the skin that is usually cooler. The temperature-sensing unit in your body is located in the hypothalamus gland. This gland is within the brain and, in reacting to changes in temperature of the blood that passes near it, sends messages throughout the body that, in turn, initiate autonomic mechanisms that act to keep the blood temperature within acceptable limits. If this regulatory process should not function correctly, major pathologies, such as hypothermia or sunstroke might result.

For example, if the hypothalamus gland detects a decrease of temperature, blood vessels that lie just under the surface of skin will constrict. Less blood flows to the skin, less thermal energy is delivered to the skin (and hence lost from the body), and correspondingly more thermal energy is delivered to the brain. The sequence of photographs shows the effect of a cold environment on a man's body. As the temperature of the environment decreases, blood vessels under the skin constrict, less thermal energy is brought to the skin, and the temperature of that region drops. As we shall see later, when we discuss radiation, this will cause a decrease in the amount of infrared radiation that is emitted from the body. These photographs, which were made with an infrared sensing device, show how the cold (darker) areas gradually grow.

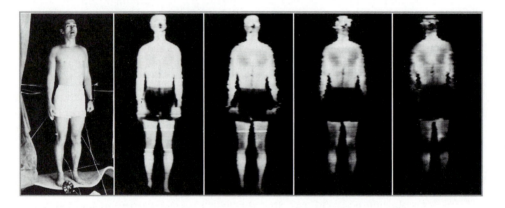

If, on the other hand, the internal temperature of the body should increase by as little as 0.5°C, these same blood vessels will dilate such that the blood flow will increase by a factor of 7. (Eckert and Randall, 1983, p. 133). Thus thermal energy will be delivered to the skin more rapidly and may then leave the body faster. There are many other such temperature regulatory mechanisms, to be discussed later. Temperature regulation is but one example of your body's need to maintain a relatively constant internal environment (**homeostasis**). This need is so important that the body can actually cut off blood circulation to various parts to maintain the temperature of the blood circulating through the brain. This reduction of circulation may lead to frostbite, and possibly gangrene, of the nose, fingers, toes, and other extremities.

A very interesting example of your body's ability to regulate blood circulation to stabilize the temperature of your brain is shown in the figure on page 185 (Falk, 1993). The brain generates a great deal of thermal energy as a result of its neural activity. This thermal energy must be removed, or the temperature will increase. As blood circulates through the brain, it picks up thermal energy and then leaves via the system of veins in your head and neck. However, there are two such exit routes. One passes through the external jugular vein and the other through the internal jugular vein. Venous blood from the brain may leave via either route. If the body becomes overheated, blood vessels just below the surface of the face dilate, blood flow through them increases, and thermal energy is trans-

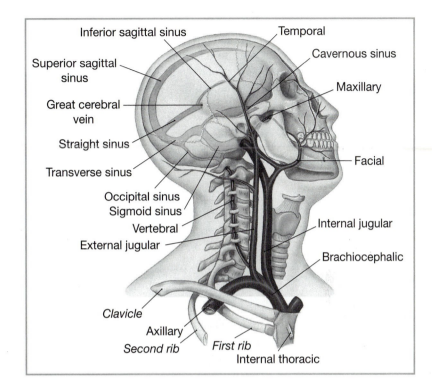

ferred from them through the skin to the environment. Venous blood from the brain passes near these vessels and gives up thermal energy to them. This cooled venous blood could return to the trunk of the body through the external jugular vein, but it does not. It reenters the brain, picking up more thermal energy, and then exits from the head via the internal jugular vein.

Convection dominates, not only within the body, but also in the loss of thermal energy to the environment. You are aware that you can feel chilly when exposed to wind, even on a warm day. If you ride a bicycle, you may feel fine as long as you are moving but may feel very hot if you come to a stop. If you blow on your hand, it will feel cool, even though your breath is warm. These are all examples of convection of thermal energy from your body to the environment via moving air. This is a major factor in the thermal balance of your body.

The convection of thermal energy between an object and its environment may be expressed by the following formula:

$$\frac{TE_{convection}}{\Delta t} = K_{convection} A_{surface} \left(T_{object} - T_{environment} \right)$$

where $A_{surface}$ is the surface area of the object over which the moving medium passes.

In dealing with the human body, it is useful to know that a reasonable value for its surface area is 1.5 m^2. In any analysis, it is necessary to estimate the fraction of this area that is exposed to the moving air.

We have already pointed out that the rate of thermal energy convection depends on the speed of the moving medium; this is built into the coefficient, $K_{convection}$. This coefficient is extremely difficult to discuss from the point of view of basic theory. However, it can be represented by the following experimentally determined expression (Cameron, Skofronick, and Grant, 1992, p.86):

$$K_{convection} = 10.45 - v + 10\sqrt{v} \; \frac{kcal}{m^2 \cdot h \cdot °C} \qquad \text{for } 2\,\frac{m}{s} \leq v \leq 20\,\frac{m}{s}$$

EXAMPLE 3.22

Let us assume that the woman in Example 3.21 has 0.6 m^2 of skin exposed to the air. How fast would air (assumed to be at 70°F) have to be moving past her for her skin temperature to stay at 90°F?

Solution

We can assume that for her skin to stay at a constant temperature, thermal energy must be leaving it at the same rate as it is being produced. We have already shown that her body is producing thermal energy at the rate of 153 kcal/h. We can substitute this into the equation that describes convection of thermal energy:

$$\frac{TE_{convection}}{\Delta t} = K_{convection} A(T_{person} - T_{environment})$$

$$153\,\frac{kcal}{h} = K_{convection}(0.6\text{ m}^2)(32.2°C - 21.1°C)$$

Thus $K_{convection}$ equals 23 kcal/(m^2 °C h). The following formula relates the convection coefficient to the velocity:

$$K_{convection} = 10.45 - v + 10\sqrt{v}$$
$$23 = 10.45 - v + 10\sqrt{v}$$

This equation may be solved graphically or algebraically and yields

$$v = 2.1 \text{m/s} = 4.7 \text{ mi/h}$$

This is quite a reasonable speed for bicycle riding. She will not feel overheated as long as she keeps moving. If she stops, however, the convection loss will decrease dramatically, and her internal temperature will rise.

EXAMPLE 3.23

Consider a person with an exposed area 0.8 m^2 and a skin temperature of 30°C who is sitting still while exposed to a 20 mi/h wind that is at a temperature of 21°C. What will be the effect on the person?

Solution

The coefficient of heat convection will equal 31.4 kcal/(m² °C h). He will be losing heat because of convection at the rate of 226 kcal/h. Since his body is producing heat at the rate of 86 kcal/h (his BMR), there is a net loss of internal heat, and his body temperature will fall. This will lead to hypothermia.

Conduction is thermal energy transfer that is accomplished via a medium, but in this case the medium does not move. Common examples of this are thermal energy transfer through the walls of a building, through a person's skin or through clothes. This process can be described by the following equation:

$$\frac{TE_{conducted}}{\Delta t} = K_{conduction} A_{surface} \frac{\Delta T}{D}$$

where Δt is the amount of time during which the transfer occurs, $A_{surface}$ is the area of the surface across which the transfer occurs, D is the length of the conducting path, $K_{conduction}$ is the coefficient of thermal conductivity, and ΔT is the difference in temperatures of the two surfaces of the conducting material.

MATERIAL	COEFFICIENT OF THERMAL CONDUCTIVITY $(K_{conduction})\ \dfrac{J}{m \cdot s \cdot °C}$
Silver	406
Copper	385
Aluminum	205
Steel	50.2
Ice	1.6
Human skin	0.84
Concrete	0.8
Glass	0.8
Water	0.67
Red brick	0.6
Snow	0.42
Body fat	0.17
Insulating brick	0.15
Wood	0.12–0.04
Dry beaver fur	0.05
Caribou fur	0.04
Felt	0.04
Glass wool	0.04
Down	0.03
Air	0.024
Styrofoam	0.01
Tissue (unperfused)	18 kcal cm/(m² h°C) (Davidovits, 1975, p. 111)

In dealing with building insulation, the formula given above for the conduction of *TE* is not used. The following formula is used:

$$\frac{TE_{\text{conduction}}}{D\Delta t} = \frac{A\Delta T}{R} \quad \text{BTU per hour per foot thickness of insulation}$$

where A is measured ft^2 and ΔT in °F.

The thermal conductivity of the body (assuming no flow of blood) is very low, comparable to that of cork. Given that $K_{\text{body tissue}}$ is approximately 18 kcal cm/(m² h°C) and assuming a thickness of 3 cm, an area of 1.5 m², and a difference in temperature of 2°C, we get a thermal energy flow of 18 kcal/h. To increase this flow to 153 kcal/h (referring again to the bicycle-riding woman), we would require a temperature difference of 17°C (the temperature of the environment would have to be 20°C (68°F)). This situation cannot be counted on. The thermal conduction of air is even smaller: K_{air} is approximately equal to $K_{\text{tissue}}/9$. The very low thermal conductivity of the body explains why it is the circulation of the blood that provides the major means of thermal energy transfer within the body. **Thus, the circulatory system is not only responsible for supplying cellular oxygen and nutrition and the removal of cellular waste, but also is vitally responsible for the movement of thermal energy around the body.**

Notice that metals have large values of thermal conductivity. This means that they are good conductors of thermal energy. The nonmetals have low values; they are poor conductors of thermal energy. These same statements could be made about electrical conduction. So it is not a surprise to learn that the main process (on the atomic scale) that serves to conduct thermal energy also conducts electricity. Note the coefficient for air. It is extremely low. Air should be a good thermal energy insulator. Why, aside from aesthetic considerations, can we not walk around naked in the winter? Air may not serve well to conduct thermal energy, but it serves very well to support convection. For us to keep warm, it is necessary that we trap still air near our bodies. This is what clothes do. Why do we get gooseflesh when we get cold? If you watch a cat when it is cold, its hairs stand up to trap air as an insulating layer. We humans are trying to do the same thing. We have lost most of our body hair, but the muscles that made the hairs stand up are still active. It is the pull of these muscles that makes our skin pucker into gooseflesh.

As we can see by looking at the basic equation that describes thermal energy conduction, the quantities ΔT and D are very important. The rate at which thermal energy is being conducted between two regions depends directly on the difference between the temperatures of the two regions and inversely on the distance between them. This means that the greater the difference between the two temperatures, the faster will be the resulting rate of thermal energy conduction between the two regions.

So, for example, thermal energy would be conducted from a penguin's feet to the ice at a high rate, as would thermal energy be conducted from a tuna fish's body to the cold water in which it is swimming. Such animals (both cold-blooded and warm-blooded) have a *rete mirabilis*, a region where arterial blood passes close to venous blood in a structured manner to control thermal energy transfer. If the arterial blood were to go directly from the core of the body, where much of the thermal energy is generated, directly to the feet or skin, much of its thermal energy would be lost from the body. The relatively hot arterial blood would be in close proximity to the relatively cold environment. The body would thus lose a significant amount of thermal energy. Since the venous blood has already lost some of its thermal energy as it circulated near the skin or through the feet, its temperature is lower than that of the arterial blood coming from the heart. Thus the difference in temperatures is large, and the distance between the two regions is small. This leads to a high rate of conduction of thermal energy from the arterial blood to the venous blood. The venous blood now returns the thermal energy to the core of the body rather than have it leave the body through the skin or feet.

There is another, more specific kind of a rete, called a *carotid rete*. This has the specific function of controlling blood circulation to moderate any increase of the temperature of the brain. Humans do not have such a rete, but there are two groups of animals that do (Baker, 1993). They are the artiodactyls (even-toed hoofed mammals, such as antelope, camels, sheep, goats, and cows) and carnivores (such as seals, sea lions, and wild and domestic cats and dogs). It is important to note that these animals either do not sweat or can avoid sweating even when exerting themselves. They do, however, pant. These animals' muzzles have a bony network through which venous blood circulates just below the surface, from which water evaporates as the animal pants. The blood is thus cooled, transferring thermal energy to the evaporating water. The cooled venous blood then circulates through the cavernous sinus where the carotid rete is located. Here, the cool blood passes near to the arterial blood going to the brain. Thermal energy from the hot arterial blood transfers to the cool venous blood, and therefore the brain is protected from overheating (Eckert and Randall, 1983, p. 138 and Irving, 1966, p. 101.)

Radiation is the transfer of thermal energy without requiring a medium. The most common example of this is the transfer of thermal energy from the sun to the earth. This thermal energy gets to us through the emptiness of space. We also receive light from the sun. It also gets to us through empty space. This is no coincidence. Light is one form of radiated thermal energy. From experiments, we know that when an object gets hot enough, it will radiate light. The term "hot enough" makes the link with thermal energy. Since the energy that is being radiated happens to stimulate our visual system, we call it *light*. If an object is too cold to radiate light, it may still be radiating something—we just can't see it (it doesn't stimulate our visual system). Such radiation is called *infrared*.

The surprising thing is that we ourselves are radiating energy. In fact, a large fraction of the thermal energy that we lose to keep our body temperature within bounds is in the form of radiation.

There are two equations that are used to analyze this type of thermal energy transfer. The net rate at which an object is emitting thermal radiation is given by the Stefan-Boltzmann law:

$$\frac{TE_{radiated}}{\Delta t} = A\varepsilon\sigma\left(T_{object}^4 - T_{environment}^4\right)$$

where σ is the Stefan-Boltzmann constant, ε is the emissivity of the surface, A is the surface area in square meters, and T is the absolute temperature in Kelvin. The net rate is used because the object is not only emitting radiation, it is also absorbing radiation that is emitted from the environment.

The equation shows a relation between the net rate of emission of thermal energy and the temperatures of both the object and its environment. Notice that the rate of emission (closely related to the intensity) of the radiation depends on the absolute temperatures raised to the fourth power. A reasonable figure for the net rate of radiated energy for a person is approximately 100 W (86 kcal/h).

Since, in most cases involving the human body, the temperature of the body is not too different from the temperature of the environment, the following form of the Stefan-Boltzmann law may be used:

$$\frac{TE_{radiated}}{\Delta t} = K_{radiation} A (T_{body} - T_{environment})$$

where $K_{radiation}$ is approximately 6 kcal/(m² h°C).

EXAMPLE 3.24

Consider once again the woman who is riding the bicycle (Example 3.21). We already know that she is generating thermal energy at the rate of 153 kcal/h and losing thermal energy through convection at the same rate. Now assume that she is riding a stationary bicycle, perhaps in a gym where the temperature is 68°F (= 20°C). Now there will be no loss of thermal energy due to convection. Can radiation keep her temperature stable?

Solution

$$\frac{TE_{radiated}}{\Delta t} \approx (6)(0.6)(303 - 294)$$

$$\frac{TE_{radiated}}{\Delta t} \approx 32 \ \frac{kcal}{h}$$

This is only about 20% of the rate at which she is generating thermal energy, and therefore radiation alone will not keep her body at a stable temperature.

When a person is exposed to bright sunlight, the previous equations will not apply. The temperature of the environment is not the important factor, since the radiation is coming from the sun, not from the immediate environment. Under this condition, the following equation gives the rate of thermal energy transfer to the body (Davidovits, 1975, p. 142)

$$\frac{TE_{radiation}}{\Delta t} = 575 \varepsilon A \cos \theta \qquad \frac{kcal}{h}$$

where ε of the body is typically about 0.6 and θ is the angle at which the sunlight is hitting the body (which may be taken to be 60° at the latitudes associated with North America, Europe, and Asia).

EXAMPLE 3.25

Refer again to the woman riding the bike (Example 3.21). If she were to be riding in bright sunshine, she would be gaining thermal energy through radiation at the rate of

$$\frac{TE_{\text{radiation}}}{\Delta t} = (575)(0.6)(0.6)(\cos 60) \qquad \frac{\text{kcal}}{\text{h}}$$

$$\frac{TE_{\text{radiation}}}{\Delta t} = 104 \qquad \frac{\text{kcal}}{\text{h}}$$

Thus her body's thermal energy would be increasing at the rate of 257 kcal/h. **Convection at 4.7 mi/h would not remove thermal energy rapidly enough, and her internal temperature would rise.**

EXAMPLE 3.26

Once again, consider the bicycle-riding woman. When we left her (Example 3.25), she was riding her bike at 4.7 mi/h, generating thermal energy at the rate of 153 kcal/h, losing thermal energy via convection at the rate of 153 kcal/h, and gaining thermal energy via radiation from the sun at the rate of 104 kcal/h. At what rate would she have to be evaporating sweat so that her temperature does not increase?

Solution

The amount of thermal energy involved in a change of state is given by

$$TE_L = (H_L)(M)$$

The rate at which this thermal energy is used would be given by

$$\frac{TE_L}{\Delta t} = H_L \frac{M}{\Delta t}$$

Solving for $M/\Delta t$, the rate at which sweat is evaporated, we get

$$\frac{M}{\Delta t} = \frac{104 \, \dfrac{\text{kcal}}{\text{h}}}{\left(2428 \, \dfrac{\text{J}}{\text{g}}\right)\left(\dfrac{1 \, \text{kcal}}{4187 \, \text{J}}\right)}$$

She would have to sweat at the rate of 179 g/h or about 0.4 lb/h to keep her temperature from rising. Although this might seem high, it is a quite reasonable rate of water loss while bicycling in bright sunshine.

The Wien equation shows a relationship between the temperature of an object and the wavelength (λ) of the most intense radiation that it emits. It is this equation that represents the change in color of the light emitted by a nail as it is held in a fire and gradually heats up. The nail is initially invisible (assuming no ambient light); it starts glowing a dull red, becomes more orange, then yellow, and finally white. This progression of colors from red to yellow represents the change in wavelength expressed by the Wien equation:

$$\lambda_{max}T = 2.898$$

where λ_{max} is the wavelength of the most intense radiation and is measured in millimeters. T is measured in Kelvin. This equation can be used to determine the temperature of an object if we know the wavelength (color) of the most intense light that it emits. Conversely, if we know the temperature of the object, we can determine the wavelength (color) of the most intense light that it emits.

EXAMPLE 3.27

The temperature of a person's skin is 93°F. What is the color of the light that he is emitting?

Solution

We can use the Wien equation to calculate λ_{max}. The skin temperature (93°F) must be converted to Kelvin (307 K) $\lambda_{max} = 9.4 \times 10^{-6}$ m. If we refer to the solar emission curve, we see that the light emitted by the person does not fall within the region that can be seen by the human eye. It corresponds to a shorter wavelength and is therefore in the region called infrared (IR).

The rods and cones in our eyes do not react to IR light, and so, you cannot see another person in the absence of sufficient ambient light. However, we are emitting IR light and there are biologic, photographic, and electronic detectors that will react to this radiation. There is a rattlesnake that lives in the southwest desert of the United States, called the sidewinder. It has a special organ in pits on either side of its head that reacts to thermal radiation in the IR region. The snake uses this to detect the presence and location of mammals that are abroad on the desert at night. Since there is little or no water on the desert, little thermal energy is stored during the day, and at night temperatures fall quickly, except for those of mammals, whose body temperatures must stay up. So they represent localized hot spots on a cold background. They emit much more IR radiation than does the colder background. The snake detects this difference and thus can hunt them.

A person's body is not at a uniform temperature. There are localized hot areas and cold areas. The hotter areas are usually those where there is more blood circulation. The differences in temperatures result in differences in IR radiation. There is an imaging process called *thermography* that detects the variations in IR radiation and produces images of them. During the 1960s and early 1970s, thermography was forecast to become a reliable technique for mass noninvasive screening for cancer and problems related to blood circulation, such as stroke. Unfortunately, the technique has not lived up to its promise.

The pictures on p. 193 were taken by using IR emission from a person's arms (Gershon-Cohen, 1967). The picture on the left was taken before the person had smoked a cigarette, and the one on the right was taken 15 minutes afterward. The decrease in IR emission is clear. This is explained by the vascular constriction associated with intake of nicotine.

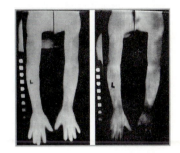

This technique may also be used to indicate the effect of a therapeutic intervention. For example, the pictures show the effect of surgery that corrected a blockage of the carotid artery (Gershon-Cohen, 1967). Notice that in the left picture below, the left temple is darker (cooler) than the right one. In the right picture, both temples look the same, and, in fact, both are brighter than in the first picture. This indicates that the blockage has been removed and that blood circulation has correspondingly increased. (*Note:* There are other techniques, (such as MRI), that are used for the same analysis. The discussion here is intended to deal only with thermography.)

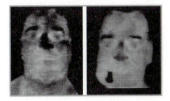

One final example of the medical use of the body's thermal radiation is shown below. This is a thermogram of a person who has a thyroid tumor (Gershon-Cohen, 1967). It is shown by the bright area below the throat.

PROBLEM SET 10

10.1. A glass coffeepot has a bottom with an area of 400 cm^2 and is 0.75 cm thick. The coffee in the pot is kept at a constant temperature of 100°C.

    a. How hot must the underside of the bottom of the coffeepot be kept so that thermal energy is transferred into the coffee at the rate of 500 W? (212°C)

    b. At what rate will coffee boil away if evaporation is the only way in which thermal energy can leave the pot? (0.22 g/s)

10.2. What is the net rate of loss of thermal energy by radiation from a flat roof (surface area = 250 m^2) if its temperature is kept at 20°C and the temperature of the environment is 10°C? (13.5 kW)

10.3. Consider a 190-lb person, nearly nude, who is walking in bright sunlight at a slow rate (3 mi/h). Assume that the air temperature is 47°C and that the person's skin is at 36°C and has an effective surface area of 1.7 m^2.

    a. At what rate is the person gaining thermal energy from radiation from the sun? (294 kcal/h)

    b. Is this person gaining or losing thermal energy by radiation exchange with the environment? How much? (gaining, 99 kcal/h)

    c. At what rate is this person gaining thermal energy via convection? (243.7 kcal/h)

    d. At what rate is this person consuming stored energy? (228 kcal/h)

    e. At what rate is thermal energy building up in this person's body? (865 kcal/h)

    f. If all of this thermal energy were to remain within the body, what would be the rate of temperature rise? (0.2°C/min)

    g. At what rate would this person have to evaporate sweat to maintain a stable temperature? (1.5 L/h)

10.4. At what rate would our bicycle-riding woman have to be sweating if she were riding a stationary bicycle at the same rate in still air in bright sunshine? (1 lb/h)

10.5. Before the widespread use of refrigerators, people stored perishable foods in iceboxes. These were basically wooden boxes with a place for a block of ice and sections for milk, vegetables, and so on. Of course, the ice would melt, and the resulting water had to be caught in a pan and then dumped out. People found that if they wrapped the ice in newspaper, it would last much longer, melting more slowly. However, the perishable food then went bad much faster.

    a. Why did the ice melt more slowly when it was wrapped in newspaper?

    b. Why did the food spoil more quickly when the ice was wrapped in newspaper?

10.6. Explain why blowing on a burnt finger makes it feel cooler.

10.7. Explain why it is so important for fur-bearing animals to preen themselves, that is, keep their fur clean.

10.8. The sun emits light over a broad range of wavelengths. The color that corresponds to the maximum intensity of sunlight is yellow (a wavelength of 5×10^{-7}m). Determine the temperature of the surface of the sun. (5.8×10^3 K)

APPENDICES

APPENDIX 1: CONVERSION FACTORS

Length: 1 m = 39.37 in. = 3.281 ft = 6.214×10^{-4} mi

Volume: 1 m^3 = 1000 L = 35.31 ft^3 = 264.2 gal

Mass: 1 Kg = 0.069 sl, also is equivalent to 35.27 oz, 2.21 lb

Force: 1 N = 0.225 lb

Energy and work: 1 J = 0.738 lb ft = 0.000948 BTU = 0.239 cal = 0.000239
Cal = 2.78×10^{-7} kWh

Power: 1 W = 0.738 lb ft/s = 0.00134 hp

APPENDIX 2: SCIENTIFIC NOTATION

It is very common, in scientific matters, to deal with very large and very small numbers. For example, a reasonable value for the diameter of an atom is 0.00000000015 meter. This way of writing the number is not widely used because it is so easy to either drop or add a zero when copying it from one place to another (as with a calculator). It is more commonly written as 1.5×10^{-10} meters. Notice that the exponent (-10) indicates the number of places that the decimal point must be moved (10) and the direction in which it must be moved (the minus sign means "to the left"). In an electricity course, you may learn that a current of 1 amp is equivalent to 6,242,200,000,000,000 electrons are passing through the wire each second. This number is much more easily (also with less chance of error) written as 6.2422×10^{15} electrons. Note that in this case the exponent tells us to move the decimal point 15 places to the right.

APPENDIX 3: SIGNIFICANT FIGURES

Another advantage of using scientific notation is that it makes it easier to correctly represent the precision with which a quantity is numerically described. For example, suppose that a person's weight is measured on a bathroom scale to be 125 lb. If we convert this to grams, we get 56561 g. This is misleading because the scale measurement shows that the weight is somewhere between 124 lb and 126 lb, near the middle of the range, perhaps within half a pound of 125 pounds. Therefore there is an implied inaccuracy of 1/2 lb (or 226 g). This means that the last three digits in the number 56561 are misleading. The best that we could say about the mass of the person is that it is somewhere between 564489 g and 56674 g.

This potential error can be avoided if we agree to use scientific notation. Now the person's weight would be noted as 0.125×10^3 pounds. The three digits (125) that appear after the decimal point are called *significant figures*. They are the numbers in which we have confidence; the last one (5) may be in error. We would say that the person's weight is known to three significant figures. When this is converted to grams, the number of significant figures cannot increase. The person's mass would be 0.566×10^5 g. Once again, the last digit (6) may be in error.

In general, the number of significant figures in the result of a calculation cannot be greater than the smallest number of significant figures in any of the quantities that were used in the calculation.

APPENDIX 4: MATHEMATICAL PREFIXES

LETTER	EQUIVALENT	EXAMPLE
f	10^{-15}	The diameter of a proton is ~1 fm $= 1 \times 10^{-15}$ m $= 0.000000000000001$ m.
p	10^{-12}	
n	10^{-9}	The diameter of an atom is ~ 0.1 nm $= 0.1 \times 10^{-9}$ m $= 0.0000000001$ m. A cell membrane is 6–10 nm thick.
μ	10^{-6}	The diameters of the largest axons are ~ 4–20 μm.
m	10^{-3}	An action potential lasts ~0.5 ms in an axon but ~275 ms in cardiac muscle.
c	10^{-2}	
k	10^{3}	A 110-lb person has a mass of 50 kg.
M	10^{6}	The diameter of the earth is ~10 Mm $= 10 \times 10^6$ m $= 10,000,000$ m.
G	10^{9}	The diameter of the sun is ~1 Gm $= 1 \times 10^9$ m $= 1,000,000,000$ m.

APPENDIX 5: SOLVING WORD PROBLEMS

There is no one way of solving word problems. As you solve more and more of them, you will develop an intuition based on experience that will lead you through to the solution. First, it is necessary to gain the experience. The following steps may be helpful to you as you go through the problems.

1. Read the problem carefully. Underline or highlight the pertinent information.
2. Look at the question. Underline or highlight the question.
3. Decide what kind of problem it is. For example, is it a Second Law problem, a conservation of energy problem, a heat transfer problem?
4. Write down the basic equation(s) for this type of problem.
5. Write down what you are going to solve for.
6. Assign an appropriate letter(s) to this word(s).
7. Make a sketch when appropriate.
8. Assign appropriate letters to the various pieces of information.

9. Pick a system of units, either SI or USA.
10. Convert units so that all given information is in the system that was picked.
11. Substitute values into the equation(s).
12. Use algebra to solve for the unknown(s).
13. Assign units to the answer(s).
14. Check the answer(s) for reasonableness.

APPENDIX 6: ALGEBRA

Solving a single linear equation:

1. The object is to manipulate the equation until the unknown quantity (usually a symbol such as x) is alone on one side of the equation.
2. Manipulation of an equation means that you can:
   a. add or subtract **the same quantity on both sides.**
   b. multiply or divide **both sides by the same quantity.**

EXAMPLE:

$$7 - 2x = 5x + 3$$

Add $2x$ to both sides:

$$7 = 7x + 3$$

Subtract 3 from both sides:

$$4 = 7x$$

Divide both sides by 7:

$$\frac{4}{7} = x$$

Solving a system of two or three linear equations:

1. Write them so that like variables (e.g., x's, y's and z's) are all in the same column and any constant term (one with no letter) is on the right-hand side of the equation.
2. Pick one of the equations. (Any one will do, but maybe one looks easier than the others.)
3. Solve the equation for one of the variables. (It doesn't make any difference which variable you pick.)
4. Substitute the resulting expression for that variable in each of the other equations. You should now have reduced the number of equations by one, and these equations should involve one less variable. So if you started with three equations

involving three variables, you should now have two equations involving two variables.

5. Repeat this process until you have only one equation that involves only one variable.

6. Solve this equation; that is, use the equation to evaluate the variable.

7. Repeat steps 4–6 and thus evaluate another variable.

8. Continue this process until all of the variables have been evaluated.

EXAMPLE:

(1) $$2 - F_1 - 3F_2 + 4F_3 = 0$$
(2) $$2F_2 + F_1 + 1 = 2F_3$$
(3) $$F_3 + 3F_1 - 5 = F_2$$

Rewrite, arranging variables in same columns and constant terms on right-hand side:

(4) $$-F_1 - 3F_2 + 4F_3 = -2$$
(5) $$F_1 + 2F_2 - 2F_3 = -1$$
(6) $$3F_1 - F_2 + F_3 = 5$$

Pick an equation (6), and solve for one of the variables, F_1:

(7) $$F_1 = \frac{5 + F_2 - F_3}{3}$$

Substitute this expression into the other two equations, (4) and (5):

(8) $$-\frac{5 + F_2 - F_3}{3} - 3F_2 + 4F_3 = -2$$

(9) $$\frac{5 + F_2 - F_3}{3} + 2F_2 - 2F_3 = -1$$

Terms can be combined so that the two equations become

(10) $$-10F_2 + 13F_3 = -1$$
(11) $$7F_2 - 7F_3 = -8$$

We now have two equations, (10) and (11), involving two variables. Solve (11) for F_2:

(12) $$F_2 = \frac{-8 + 7F_3}{7}$$

Substitute this expression into (10):

(13) $$-10\left(\frac{-8 + 7F_3}{7}\right) + 13F_3 = -1$$

We now have one equation, involving one variable. F_3 may be evaluated:

(14) $$80 - 70F_3 + 91F_3 = -7$$

$$80 + 21F_3 = -7$$
$$21F_3 = -87$$

(15)
$$F_3 = \frac{-87}{21} = -4.14$$

This value may be substituted into (12) to evaluate F_2:

$$F_2 = -5.28$$

The values of F_2 and F_3 may now be substituted into (7) to evaluate F_1:

$$F_1 = 1.29$$

There are many other techniques that may be used to solve a system of linear equations. Some calculators (e.g., the TI-85) have built-in programs that will solve such systems.

APPENDIX 7: TRIGONOMETRY

Consider the triangle shown in the figure.

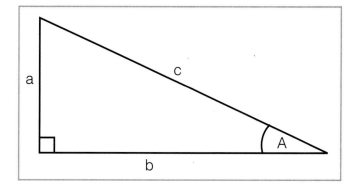

a. The sine of an angle is defined as the ratio of the opposite side to the hypotenuse:

$$\sin A = \frac{a}{c}$$

b. The cosine of an angle is defined at the ratio of the adjacent side to the hypotenuse:

$$\cos A = \frac{b}{c}$$

c. The tangent of an angle is defined as the ratio of the opposite side to the adjacent side:

$$\tan A = \frac{a}{b}$$

d. The length of the hypotenuse is related to the lengths of the other two sides by the Pythagorean theorem:

$$c = \sqrt{(a^2 + b^2)}$$

e. The size of an angle is related to the lengths of the sides of the triangle by the inverse trigonometric functions: $\sin^{-1}$, $\cos^{-1}$, and $\tan^{-1}$:

$$A = \tan^{-1}\left(\frac{a}{b}\right)$$

$$A = \sin^{-1}\left(\frac{a}{c}\right)$$

$$A = \cos^{-1}\left(\frac{b}{c}\right)$$

APPENDIX 8: REASONABLE VALUES FOR PHYSICAL QUANTITIES, AS USED IN THE TEXT

Surface area of the body	1.5 m^2
Emissivity (ε) of the body	0.6
Specific heat of the body	0.83
Earth's radius	$6.44 \times 10^6 \text{ m}$
Earth's mass	$6 \times 10^{24} \text{ kg}$
Universal gravitation constant (G)	$6.673 \times 10^{-11} \text{ SI}$

APPENDIX 9: ANATOMY

Although this book emphasizes physics, it does deal with applications that focus attention on the human body. It is therefore necessary to deal with the descriptive vocabulary, structure, and applicable geometry of the human body (see figure on page 201). We shall be using as examples the major joints of the body: tibia/talus (ankle), tibia/femur/patella (knee), femur/ilium (acetabulum) (hip), lumbar vertebrae/sacrum (lower back), radius/humerus/ulna (elbow), and humerus/scapula (shoulder). It must be emphasized that this discussion is not intended to be complete but sufficient only for the applications under consideration.

Reasonable Values for the Lengths of Major Bones

Tibia: 36 cm

Femur (greater trochanter to knee): 40 cm

Ulna: 24 cm

Radius: 22 cm

Humerus: 30 cm

trunk (shoulder to hip) (T1 to L5): 40 cm

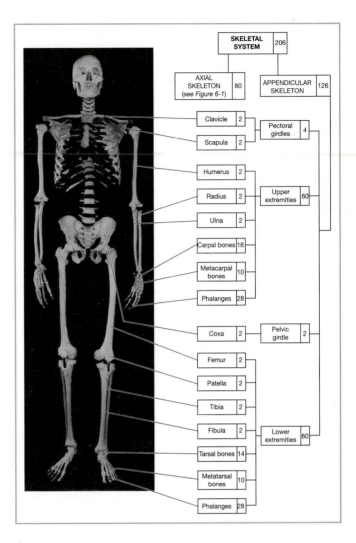

Major Joints

COMMON NAME	BONES	ABDUCTOR/ EXTENSOR	ADDUCTOR/ FLEXOR
Ankle	Tibia/talus	Achilles' tendon	
Knee	Femur/patella/tibia	Quadriceps	Hamstring
Hip	Femur/ilium	Gluteus medius and minimus	
Lower back	Fifth lumbar/ first sacrum	Erector spinae	
Shoulder	Humerus/scapula	Deltoid	
Elbow	Radius/ulna/humerus	Triceps	Biceps

Ankle (Tibia/Talus)

The ankle is a relatively weak joint, composed of the rounded ends of the tibia and talus butting up against each other. This structure, coupled with its role as supporting virtually the entire body weight and the very large forces that occur during running and jumping,

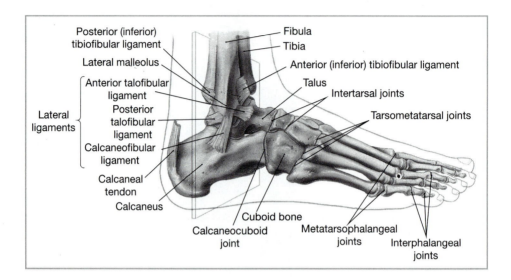

leads to many instances of dislocation and stress fracture. The ligament structure that stabilizes the ankle is easily overloaded, leading to a twisted ankle. Tension in the Achilles' tendon, originating at the calf muscle and passing parallel to the tibia, causes rotation of the foot around the ankle, but its major role is, by pulling down and backward on the tibia, maintaining the body in a vertical posture. Therefore we would expect very large amounts of stress in the Achilles' tendon when a person leans forward, particularly if the knees are not flexed.

> Geometry
> distances from ball of the foot, measured along axis of foot, to:
> talus/tibia contact: 12 cm
> heel (insertion point of Achilles' tendon): 16 cm

Knee (Tibia/Femur/Patella)

The knee, a hinge joint, is composed of three bones: the femur, tibia and patella. The patella (or kneecap) is a sesamoid bone, that is, a bone that forms within a tendon (in this case the quadriceps tendon) (see page 203). This tendon originates in the quadriceps muscle (located on the front of the thigh) and inserts onto the front of the tibia at a point that is slightly below the top of the tibia. A hinge joint is inherently less stable than a ball-and-socket joint. It is therefore much more dependent on soft tissue (ligaments) for stabilization. Within the knee, the anterior cruciate ligament (ACL), which connects the rear of the femur to the front of the tibia, must withstand great stress and often presents a potential problem for athletes. Extension of the knee is caused by the quadriceps muscle applying

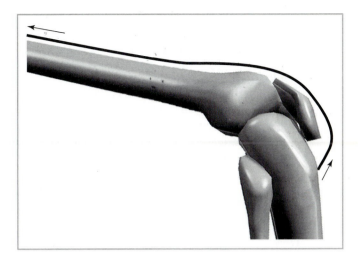

tension to the quadriceps tendon (also called the transpatellar tendon), which after passing over the patella, inserts onto the tibia. Flexion of the knee is caused by the hamstring muscle/tendon.

> Geometry
>> femur/tibia contact to front of tibia: 4 cm
>> upper end of tibia to quadriceps tendon insertion: 4 cm
>> angle between quadriceps tendon (below patella) and axis of tibia: 5°
>> quadriceps tendon (above patella) is parallel to femur

Hip (Femur/Acetabulum)

The hip joint is an example of a ball-and-socket joint. The ball (the head of the femur) fits into the socket (the acetabulum). This comprises an extremely stable joint, more subject to fracture than to dislocation. The hip abductor muscles (originating on the ilium and inserting onto the greater trochanter of the femur) serve to stabilize the joint (by pulling the femur into the acetabulum) as well as providing for abduction. We see in the text that the amount of stress in these muscles can easily exceed the body weight when a person is leaning to the side.

> Geometry
>> greater trochanter to acetabulum: 6 cm
>> angle between the axis of the femur and the neck of the femur: 125°
>> angle between the axis of the neck of the femur and the hip abductor muscle: 25°

Lower Back (Lumbar Vertebrae/Sacrum)

This (L5-S1) joint is particularly prone to problems. When a person bends forward, there is a great deal of compression, which may result in distortion or possible rupture of the cushioning tissue (disk) that separates L5 from S1. This can result in pressure on the sciatic nerve, leading to numbness, pain, or loss of control over the leg. The muscle that makes the major contribution to keeping the back erect is the erector spinae muscle.

Geometry

L5-S1 to erector spinae insertion point: $\frac{2}{3}$ the length of the spinal column

angle between the axis of the spinal column and the erector spinae muscle: 12°

Elbow (Radius/Humerus/Ulna)

The elbow is a hinge joint. It does not have a significant stabilization structure, as does the hip. Therefore it is susceptible to dislocation or soft tissue damage during activities such as baseball pitching and tennis. Extension of the elbow is caused by the triceps muscle, which inserts onto the ulna behind the elbow. Flexion of the elbow is caused by the biceps muscle, which inserts onto the radius in front of the elbow.

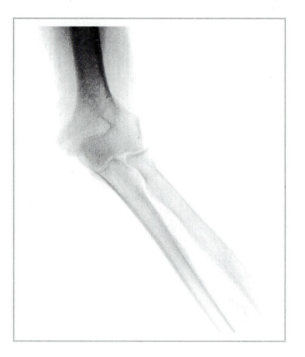

Geometry

humerus/radius contact to biceps insertion: 4 cm

humerus/ulna contact to triceps insertion: 2.5 cm

the triceps is parallel to the humerus

Shoulder (Humerus/Scapula)

The shoulder is a ball-and-socket joint, as is the hip. There is a major difference between them. While the socket (the acetabulum) of the hip is composed of bone, the socket (glenoid labrum) of the shoulder is mainly composed of soft tissue (cartilage). This makes dislocation of the shoulder much more likely than dislocation of the hip. Another difference between the shoulder and the hip is that while the hip represents almost pure rotational motion, the shoulder involves sliding as well as rotation. As the arm moves away from the body (abduction), there is rotation, but also the upper end of the humerus slides down along the scapula. This complicated motion, coupled with the lack of a bony socket structure, can lead to unacceptably large stresses on the soft tissue. These stresses, brought on by throwing, for example, may cause damage to the tissue, such as tearing of the rotator cuff. The muscle making the major contribution to eccentric motion of the arm (motion away from the body) is the deltoid muscle.

Geometry
    humerus/scapula to deltoid insertion: 13 cm
    angle between deltoid and axis of humerus: 15°

Body Segment Data

BODY SEGMENT	PERCENTAGE OF TOTAL BODY WEIGHT	LOCATION OF CENTER OF GRAVITY AS A PERCENTAGE OF DISTANCE FROM UPPER END
Head and neck	7.9	***
Trunk with head and neck	56.5	60.4
Upper arm	2.7	43.6
Forearm	1.5	43
Hand	0.6	50.6
Thigh	9.7	43.3
Lower leg	4.5	43.3
Foot	1.4	42.8

Source: Taken from *Biomechanics: Problem Solving for Functional Activity,* Roberts and Falkenburg, Mosby Year Book, 1992.

Entire arm (hand included)	4.8	46
Entire leg (foot included)	15.6	40

Estimated by author

BIBLIOGRAPHY

Adair, R.K., "Physics of Baseball," *The Physics Teacher,* May 1995, pp. 26–31.

Alexander, R.M. "Walking and Running," *American Scientist,* vol.72, July–August 1984, pp. 348–354.

Anderson, F.C. and Pandy, M.G., "Storage and Utilization of Elastic Strain Energy During Jumping," *Journal of Biomechanics,* vol. 26, no.12, 1993, pp. 1413–1427.

"An Energy-Efficient Caddy," *Discover,* July 1995, p. 18.

American Heart Association, *Exercise Standards,* February 21, 1990.

Apt, S., "No Fuel Left for LeMond," *New York Times,* December 3, 1994.

"The Bag," *Popular Science,* October 1992, p. 59.

Baker, M.A., "A Wonderful Safety Net for Mammals," *Natural History,* August 1993.

Barnes, G., "Jackrabbit Ears and Other Physics Problems," *The Physics Teacher,* March 1990, pp. 156–159.

———, "Nature's Heat Exchangers," *The Physics Teacher,* September 1991, pp. 330–333.

Benedek and Villars, *Physics,* Vol. 1, Addison Wesley, Reading, Mass., 1973.

Bennett-Clark, H.C. and Lucey, E.C.A., "The Jump of the Flea: A Study of Energetics and a Model of the Mechanism," *Journal of Experimental Biology,* vol. 47, 1967, pp. 59–76.

Blount, W.P., "Don't Throw Away the Cane," *Journal of Bone and Joint Surgery,* vol. 38-A, no. 3, June 1956.

Bronowski, J., *The Ascent of Man,* Little, Brown & Company, Boston/Toronto, 1973.

Brush, S.G., "Thermodynamic and History," *Graduate Journal,* vol. 7, no. 2, Spring 1964.

Burns, D.M. and McDonald, S.G.G. *Physics,* 2nd edition, Addison Wesley, Reading, Mass., 1975.

Cameron, J., Skofronick, J.C. and Grant, R.M. *Physics of the Body,* Medical Physics Publishing Company, 1992.

Carey, F., "Fishes with Warm Bodies," *Scientific American,* vol. 228, no. 2, February 1973, pp. 36–44.

Casper, B.M., "Galileo and the Fall of Aristotle: A Case of Historical Injustice?," *American Journal of Physics,* vol. 45, no. 4, April 1977, pp. 325–330.

Church, N.S.,"Heat Loss and the Body Temperatures of Flying Insects," *Journal of Experimental Biology,* vol. 37, 1960, pp. 171–212.

Cromer, A.H., *Physics for the Life Sciences,* McGraw Hill, New York. 1974.

Dampier, W.C., *A History of Science,* MacMillan, NY, 1944.

Davidovits, P., *Physics in Biology and Medicine,* Prentice-Hall, Englewood Cliffs, N.J., 1975.

Duncan, G., *Physics in the Life Sciences* 2nd edition, Blackwell, Cambridge, Mass., 1990.

Eckert, R. and Randall, D., *Animal Physiology: Mechanisms and Adaptations,* 2nd edition, Freeman, New York, 1983.

Enoka, R.M., *Neuromechanical Basis of Kinesiology,* 2nd edition, Human Kinetics, Champaign, IL, 1994.

Falk, D., "A Good Brain Is Hard to Cool," *Natural History,* August 1993, p. 64.

Fishbane, P.M., Gasiorowicz, S., and Thornton, S.T., *Physics for Scientists and Engineers, Extended,* 2nd edition, Prentice-Hall, Englewood Cliffs, N.J., 1996.

Fung, Y.C., *Biomechanics,* Springer, New York, 1997.

Gates, D.M., *Energy Exchange in the Biosphere,* Harper & Row, London, 1965.

Gershon-Cohen, J., "Medical Thermography," *Scientific American,* vol. 216, no. 2, February 1967.

Giancoli, D.C., *Physics: Principles with Applications,* 5th edition, Prentice-Hall, Englewood Cliffs, N.J., 1998.

Gustafson, D.R., *Physics: Health and the Human Body,* Wadsworth, New York, 1980.

Hart, S., "Caterpillar Cadillacs," *Discover,* March 1992.

Heinrich, B. and Bartholomew, G., "Temperature Control in Flying Moths," *Scientific American,* vol. 226, no. 6, June 1972, pp. 70–77.

Hobbie, R.K., *Intermediate Physics for Medicine and Biology,* John Wiley, New York, 1988.

Hoaland, C.M., "Minimum Engine Size for Optimum Automobile Acceleration," *American Journal of Physics,* vol. 60, no. 5, May 1992, pp. 415–422.

Irving, L., "Adaptations to Cold," *Scientific American,* vol. 214, no. 1, January 1966, pp. 94–101.

Jarman, M., *Examples in Quantitative Zoology,* Arnold, London, 1970.

Kluger, J., "Fairway Physics," *Discover,* August 1996, pp. 38–41.

Lenihan, J., *Human Engineering,* Braziller, New York, 1975.

Mackay, M., "Impact Biomechanics and Traffic Crashes," *Physical Medicine and Rehabilitation,* vol. 12, no. 1, 1998.

Monteith, J.L., *Principle of Environmental Physics,* Arnold, London, 1973.

Mount, L.E., *Adaptation to Thermal Environment,* Arnold, London, 1979.

Nalence, E.E., "Using Automobile Road Test Data," *The Physics Teacher,* May 1988, p. 278.

Natural History, August 1993, pp. 28–71 (collection of articles showing how various animals cope with heat).

Nicklin, R.C., "Kinematics of Tailgating," *The Physics Teacher,* February 1997, p. 78.

Nigg, B.M. and Herzog, W., *Biomechanics,* John Wiley, New York, 1995.

Norkin, C.C. and Levangie, P.K., *Joint Structure and Function,* 2nd edition, F.A. Davis, Philadelphia, 1992.

Park, J.B., *Biomaterial Science and Engineering,* Plenum, New York, 1984.

Phillipson, J., *Ecological Energetics,* Arnold, London, 1966.

Pugh, L.G.C.E., "The Influence of Wind Resistance in Running and Walking and the Efficiency of Work Against Horizontal and Vertical Forces," *Journal of Physiology (London),* vol. 213, 1971, pp. 255–276.

Rausch, P. J., *Kinesiology and Applied Anatomy,* 7th edition, Lea & Febiger, Philadelphia, 1989.

Roberts, S.L. and Falkenburg, S.A., *Biomechanics: Problem Solving for Functional Activity,* Mosby Year Book, St. Louis, 1992.

Rothwell, N.J. and Stock, M.J., "A Role for Brown Adipose Tissue in Diet Induced Thermogenesis," *Nature,* vol. 218, 1979, pp. 31–35.

Ruch T.C. and Patton H., *Physiology and Biophysics,* 19th edition, W.B. Saunders, Philadelphia, 1966.

Shames, M.H., *Great Experiments in Physics,* Holt, Rinehart and Winston, New York, 1959.

Singer, C., *A Short History of Scientific Ideas to 1900,* Oxford University Press, Oxford, England, 1959.

Singleton, W.T., *The Body at Work,* Cambridge University Press, Cambridge, England, 1982.

Smith, J.M., *Mathematical Ideas in Biology,* Cambridge University Press, Cambridge, England, 1968.

Steindler, A., *Kinesiology of the Human Body,* Charles C. Thomas, Springfield, Ill., 1970.

Tucker, V.A.,"The Energetics of Bird Flight," *Scientific American,* May 1969, pp. 70–77.

Urone, P.P., *Physics,* John Wiley, New York, 1986.

Warren, J.V., "The Physiology of the Giraffe," *Scientific American,* vol. 231, no. 5, November 1967, pp. 96–105.

Weinstock, H., "Thermodynamics of Cooling a (Live) Human Body," *American Journal of Physics,* vol. 48, no. 5, May 1980, pp. 339–341.

Wells, K. and Wessel J., *Kinesiology,* 5th edition, W.B. Saunders, Philadelphia, 1971.

Zimmer, C., "No Skycaps Needed," *Discover,* August 1995, p. 28.

———, "Beetle of Burden," *Discover,* April l996, p. 27.

INDEX